WESTERN EUROPEAN CENSUSES, 1960

UNIVERSITY OF ST. THOMAS

O'SHAUGHNESSY-FREY

LIBRARY CENTER

THIS BOOK DONATED THROUGH
THE GENEROSITY OF

The Luxemburgiana

Bach-Dunn Collection

Western European Censuses, 1960

AN ENGLISH LANGUAGE GUIDE

JUDITH BLAKE
JERRY J. DONOVAN

Institute of International Studies
University of California, Berkeley

Standard Book Number 87725-308-0
Library of Congress Card Number 77-634274
© 1971 by the Regents of the University of California

ACKNOWLEDGMENTS

This publication is part of the International Census Documents Project, Department of Demography, University of California (Berkeley). We gratefully acknowledge the support of training grants to the Department of Demography from the Ford Foundation, the National Institute of General Medical Sciences (T01 GM 01240), and the National Center for Health Services Research and Development (Public Health Service Grant No. 8 T01 HS00059).

This monograph was begun by Professor Blake during the summer of 1966 with the assistance of two students, Katherine K. Carter and Valerie A. Caires. They worked on it part time for over a year. Ann E. Norcross replaced them in the fall of 1967, and continued with the project on a full-time basis until May 1968. Mr. Donovan began working on the monograph in the summer of 1968, and has worked on it as co-author through its completion.

We are particularly indebted to the staff of the Documents Department of the University of California (Berkeley) Library: George R. Davis (Assistant Head of the Department), Sheila T. Dowd, Corinne J. Gow, and Kevin R. Gregg, who have provided invaluable technical help, encouraged us in our efforts, and borne patiently with our incessant needs and requests. As translators, John M. Hoberman, Lee M. Spangler, Irene L. Klemm, Judith Bernstein, and Regina B. Arriaga have helped us unlock the information for the use of the non-linguist. The skillful copy-editing of Paul Gilchrist and Neda A. Tomasevich has aided us greatly. Finally, we must give special thanks to Linda G. Carlson, Linda F. Baker, Maya E. Spence, and Bojana Ristich, who were responsible for an onerous typing chore.

J. B.
J. J. D.

CONTENTS

INTRODUCTION	1
AUSTRIA	8
BELGIUM	19
DENMARK	33
FINLAND	44
FRANCE	54
GERMANY	92
GIBRALTAR	123
GREAT BRITAIN [ENGLAND AND WALES]	125
GREECE	175
IRELAND (EIRE)	200
ITALY	218
LIECHTENSTEIN	245
LUXEMBURG	249
MALTA	259
THE NETHERLANDS	264
NORTHERN IRELAND	282
NORWAY	293
PORTUGAL	304
SCOTLAND	317
SPAIN	348
SWEDEN	360
SWITZERLAND	385

APPENDIXES

A.	Economic Activity	409
B.	Employment Status	411
C.	Industrial Classification	413
D.	Occupational Classification	415
E.	Socio-economic Categories	416
F.	Households and Family Nuclei	417

INDEX: SELECTED TOPICS TABULATED BY NATIONAL CENSUSES   420

INTRODUCTION

Growing scientific interest in the structure, characteristics, and dynamics of the world's population is creating an unprecedented scholarly demand for international census documents. Yet, as the researcher soon discovers, these materials are neither widely available nor readily accessible for systematic analysis. The present volume, treating the 1960 censuses of Western Europe, is an experimental effort to overcome, for the scholar who reads English, some of the most vexing problems of census availability and accessibility.[1]

Census data may be said to be <u>available</u> if the censuses are located within a geographical radius not appreciably greater than that involved in commuting. However, outside of Washington, D.C., American scholars, including those near the best-endowed of our educational institutions, do not have even approximately complete international census data available to them.[2] Away from the major centers of learning, the individual scholar must depend heavily on secondary sources (such as yearbooks) and limited interlibrary borrowing. Under such circumstances, he is rarely able to borrow an actual foreign census volume.

Even when censuses are available, they may not be <u>accessible</u> for scholarly work. If they are in many different foreign languages, for example (such as is the case with the 22 censuses of Western Europe), to obtain even a superficial notion of their contents is a major task. Is it possible to do a comparative analysis of some aspect of the labor force, educational attainment,

---

[1] The information in this volume on the location of censuses in American libraries will be of value mostly to scholars in the United States. Aside from this limitation, the volume should be of use to readers of English everywhere.

[2] Universities rarely inventory their census collections. We have finished an inventory of the population censuses of Western European nations (1960 retrospectively to the earliest enumerations) held in the collection of the University of California (Berkeley) Library, which indicates the current holdings to be 58 percent of all population censuses known to be published. The University of Texas estimates its holdings to be 50 percent of those censuses listed in its bibliographies of international censuses.

fertility, marital status, commuting, occupational structure, or one of the many other topics buried in these volumes? To answer such a question--indeed, to discover what kinds of analysis cannot be done--is in itself a research project. Moreover, the scholar must often contend with lost, misplaced, or already checked-out volumes that are only "technically" available, or possibly wait a considerable amount of time while a volume is bound and catalogued.

This book is an attempt to circumvent many of the problems arising from the unavailability and inaccessibility of censuses by (1) giving the titles (translated into English where necessary) and the page numbers of all the statistical tables in every volume of every Western European census taken during the 1960 census period;[3] (2) providing, for each census, a detailed glossary of technical terms that appear in more than one volume of that census; (3) annotating all tables having unique technical terms; (4) providing detailed appendixes on major concepts and classifications that cut across all or most of the censuses (e.g., occupational and industrial groups, the labor force, household and family structure, cultural and personal characteristics), and (5) giving a bibliographically correct entry for every volume of every census to facilitate identification when inquiries and requests to research libraries are made. One verified location for every census volume has been determined--the Documents Department of the University of California (Berkeley) Library.[4]

In sum, this volume gives the scholar a compendium in English of the contents of all the Western European censuses for the 1960 period. From it he can learn whether comparative analysis of a particular research topic (the composition of the labor force or

---

[3] The census of Monaco is not included because it apparently has not been published.

[4] For locating census volumes, the definitive list of holdings of principal research libraries in the United States and Canada is the Library of Congress's National Union Catalog. The National Union Catalog normally gives open entries (an entry for the entire census but not for individual volumes) for census publications having more than one volume per entry. The libraries reported as holding certain entries may not possess all the volumes of the census in question. It is necessary, therefore, to verify at the reporting library the completeness of its holdings. There are also regional union catalogues of library holdings in the United States, which may be found in library reference collections. The holdings of the U.S. Bureau of the Census Library are listed in U.S. Department of Commerce, Bureau of Census, Foreign Statistical Publications. Accessions List.

INTRODUCTION

fertility, for example) is possible, and he can make a preliminary decision concerning which tables in any of the 22 censuses interest him. Should he wish to order photocopies of these tables, he has been advised of at least one library where the volumes containing them may be found, and he can give exact bibliographical references for the volumes, including the page numbers of the tables in question. When he receives copies of the tables (in the native language), he is aided in using them by the glossaries, appendixes, and footnotes provided here. (If the census volumes are located in the scholar's own institutional library, his work will be facilitated by the availability in English, in one volume, of the table titles for all 22 censuses.)

There is, of course, much valuable information in census publications not contained in the principal statistical tables. For example, there are often analyses of trends, text tables, and summaries to be found in introductions to census volumes. With one or two exceptions, such materials are noted here only if they have found their way into the glossaries and appendixes; they have been given no independent recognition. Moreover, we have not included references to, or information about, supplementary volumes growing out of a census, or supplementary statistical tables that sometimes appear in the form of computer output. We also have left the table titles unedited, even though they may not, at times, completely and accurately state which variables appear in the tables. As a consequence, the English table titles for different countries are not completely consistent with each other in form and terminology--including the table titles of censuses of the English-speaking countries and of those few European countries which provided English translations.[5]

A detailed description of the major organizational foci of the volume follows.

Entries

An _entry_ for a census is the bibliographical form by which it can be uniformly identified and cited. We chose to use the U.S. Library of Congress form of citation because of its universal acceptance by U.S. research libraries, as well as its growing acceptance in libraries internationally. Where Library of

---

[5]The table titles for Great Britain, Northern Ireland, Scotland, Gibraltar, and Malta were published in English, and those for Finland, the Netherlands, Norway, and Sweden appeared in English translation along with the indigenous language. For Greece, English translations were available for the sample tabulations only.

INTRODUCTION

Congress entries exist in the <u>National Union Catalog</u> (or its predecessor catalogues of Library of Congress and other holdings), we used them as the basis for our own entries, citing the L.C. numbers in parentheses. We occasionally altered the L.C. entries when, for example, additional information permitted the closing of a previously open entry. We indicated the total number of volumes when publication of a series was completed, and also the imprint date of the final publication in the series. When a census was not yet catalogued by the Library of Congress, we developed our own entry, using L.C. style as set forth in <u>Anglo-American Cataloging Rules</u> (Chicago: American Library Association, 1967).

    Censuses are usually published in the form of monographs, rather than in serials. However, some censuses appear as serial monographs--that is, the text of the census is complete in a number of related volumes having a single title. The text of other censuses is complete in a group of multiple monographs having different titles, with no additional notation serially identifying the group. Still other censuses are textually complete in some combination of serial monographs and multiple- or single-volume monographs.

    These formats find bibliographical expression in two types of entries--<u>single</u> and <u>multiple</u>. A <u>single entry</u> citation consists of one citation which stands for:

    (a) One volume (a "monograph") which, taken alone, comprises the complete text of the census (e.g., Liechtenstein); or

    (b) A group of explicitly related volumes ("serial monographs") which, taken together, constitute the complete text of the census (e.g., Germany).

A <u>multiple entry</u> citation designates complete texts of censuses which are composed of bibliographically separate monographs (i.e., a group of volumes without explicit serial relationship). A group of multiple entries may be composed of:

    (a) Citations for multiple monographs (e.g., Malta); or

    (b) A combination of citations for single volumes or multiple monographs and serial monographs (e.g., France).

As may be seen from the following breakdown, most of the 1960 censuses of Western Europe have been published in the form of a serial monograph having a single entry.

INTRODUCTION

SINGLE ENTRIES		MULTIPLE ENTRIES	
Single Volume Monographs	Serial Monographs	Multiple Monographs	Single Volume or Multiple Monographs and Serial Monographs
Gibraltar	Austria	Malta	Denmark
Liechtenstein	Belgium		France
	Finland		Great Britain
	Germany		Greece
	Ireland (Eire)		
	Italy		
	Luxemburg		
	Netherlands		
	Northern Ireland		
	Norway		
	Portugal		
	Scotland		
	Spain		
	Sweden		
	Switzerland		

Exclusions

We have systematically excluded housing censuses and preliminary reports of population censuses.

Glossaries and Appendixes

The glossary for each census is divided into two sections. The first part gives the terms, in the indigenous language and English, of the major Administrative, Geographical, and Political Units in the country, along with a brief description of each. Also included are areas whose designations are derived from census information concerning density and propinquity--urban areas and urban agglomerations, for example.

The second part of each glossary consists of a Table of Demographic Concepts, also in both the indigenous language and English. The table is subdivided into sections such as Cultural, Educational, and Personal Characteristics, Economic Characteristics, and Household and Family Characteristics. The Table of Demographic Concepts refers the reader to a set of appendixes, which provide detailed definitions and a discussion of the demographic concepts and their relations to the recommendations of the European Conference of Statisticians concerning the 1960 censuses. These recommendations and an analysis of the extent to which they were adopted appear in United Nations, Economic Commission for Europe, Conference of European Statisticians, <u>Statistical Standards and Studies, No. 3. European Population Censuses: The 1960 Series. International Recommendations and National Practices</u> (New York, 1964). Also important are the <u>International Standard Classification of Occupations</u> (ISCO), published in 1958 by the International Labor Office, and the <u>International Standard Industrial Classification of All Economic Activities</u> (ISIC), United Nations Statistical Papers, Series M, No. 4, Rev. 1 (New York, 1958).[6]

One of our primary aims in preparing the glossaries, the translated table titles, and the footnotes to individual tables was to attempt to establish uniform English terminology in those cases where concepts expressed in a wide variety of indigenous terms actually had the same technical meaning. The reader will find, for example, that even among countries sharing the same language, the same terms are not used for the same concepts, despite the fact that the same concepts are used throughout the censuses. Austria refers to "occupational classification" as <u>beruflicher Zugehörigkeit</u>; "major groups" within this classification as <u>Berufsabteilungen</u>; "minor groups" as <u>Berufsobergruppen</u>; "unit groups" as <u>Berufsgruppen</u>; and "unit subgroups" as <u>Berufsarten</u>. Germany, on the other hand, uses somewhat different terms for the same concepts, as follows: <u>beruflicher Gliederung, Berufsabteilungen, Berufsgruppen, Berufsordingen, Berufsklassen</u>. We researched in detail such nuances of usage, which occur rather frequently, and in the glossaries standardized the foreign terms in technical English terminology. The conceptual referents of this terminology are to be found in the Appendixes. Standardization was not possible for all terms, however, particularly those referring to densely populated areas, and in such cases we simply defined what each term meant in the country in question. It is left to the researcher to compile for himself a list of those terms that are roughly similar.

An effort such as this monograph doubtless contains many errors, though we hope that they have been kept to a minimum through diligent checking and proofreading. We anticipate finding that some of the concepts in the foreign censuses have been misinterpreted. Such misinterpretations point up the need for future census compilers to develop clearer definitions. In many cases, we found conceptual distinctions ambiguous and baffling, despite the fact that, in preparing this volume, all of the explanatory material from the census volume was carefully examined for

---

[6] Unless otherwise noted, the information in the glossaries comes from the censuses themselves.

INTRODUCTION

clarification. For example, classifications of social and socio-economic status were often cumbersome and complex, and their empirical referents sometimes virtually impossible to identify. Only by making some rather farfetched metaphysical assumptions could we determine what appears to be indicated by the classification. We invite our readers to tell us of errors so that we may improve a future edition of this volume, and take their comments into account as we complete work on a similar volume for the 1950 Western European censuses.

A U S T R I A
_____

MAIN ENTRY AND VOLUME NOTES:

Austria. Statistisches Zentralamt.
   Volkszählungsergebnisse 1961. Wien, 1961-1966.
   17 v. 30cm.   (NUC66-75240)

Heft 1. Vorläufige Hauptergebnisse der Volkszählung vom 21. März 1961 nach Gemeinden. 1961. 150 p.

Heft 2. Burgenland. 1963. 69 p.

Heft 3. Vorarlberg. 1963. 60 p.

Heft 4. Tirol. 1963. 69 p.

Heft 5. Salzburg. 1963. 62 p.

Heft 6. Kärnten. 1963. 67 p.

Heft 7. Niederösterreich. 1963. 116 p.

Heft 8. Oberösterreich. 1964. 76 p.

Heft 9. Steiermark. 1964. 86 p.

Heft 10. Wien. 1964. 65 p.

Heft 11. Österreich. 1964. 60 p.

Heft 12. Die Haushalte in Österreich. 1964. 24 p.

Heft 13. Die Zusammensetzung der Wohnbevölkerung Österreichs nach allgemeinen demographischen und kulturellen Merkmalen. 1964. 94 p.

Heft 14. Die Berufstätigen nach der beruflichen Zugehörigkeit. 1964. 930 p.

Heft 15. Die Berufstätigen nach ihrer wirtschaftlichen Zugehörigkeit. 1964. 176 p.

Heft 16. Wohngemeinde- Arbeitsgemeinde der Beschäftigten in Österreich. 1965. 505 p.

Heft 17. Die Wohnbevölkerung Österreichs nach Einkommensquellen und wirschaftlicher Zugehörigkeit. 1966. 208 p.

ENGLISH TRANSLATION OF CENSUS AND VOLUME TITLES:

Results of the population census of 1961.

   1. Preliminary main results of the population census of March 21, 1961, by commune.

GLOSSARY

2. Burgenland.
3. Vorarlberg.
4. Tirol.
5. Salzburg.
6. Kärnten.
7. Niederösterreich.
8. Oberösterreich.
9. Steiermark.
10. Vienna.
11. Austria.
12. Households in Austria.
13. Composition of the Austrian resident population by general demographic and cultural characteristics.
14. Economically active population by occupational classification.
15. Economically active population by industrial classification.
16. Communes of residence--communes of work of employed persons in Austria.
17. Resident population of Austria by source of income and industrial classification.

ENUMERATION DATE: March 21, 1961.

ENUMERATION BASIS: De jure.

GLOSSARY

ADMINISTRATIVE, GEOGRAPHICAL, AND POLITICAL UNITS

Level	English Translation	Vernacular Term
National	Republic of Austria	Republik Österreich
Intermediate	Federal state	Bundesland
	Political district	Politische Bezirk
Local	Commune	Gemeinde

The Austrian government is federal in form, with the duties and responsibilities divided between the nation and the states in such a way that authority which is not expressly delegated to the central government falls to the independent states.

AUSTRIA

<u>Political districts</u>, which have no legal identity of their own, stand between communes and states in the administrative structure.

The basic units of local government are the <u>communes</u>.

<u>TABLE OF DEMOGRAPHIC CONCEPTS</u>

The demographic concepts which appear in the following table correspond to the recommendations of the European Conference of Statisticians, unless otherwise noted. They are defined and discussed in the appendixes and in footnotes. Terms in parentheses are alternative forms which appear in the table titles. Terms in brackets are not used in the table titles, but they are included here in order to give logical completeness to the various classifications in the table.

Appendix	English Table Term	Vernacular Table Term
	ECONOMIC CHARACTERISTICS	
A	Economic activity	Berufskennzeichnung
	Economically active	Berufstätige
	Employed	Beschäftigte
	Unemployed	Arbeitslose
	[Economically inactive]	...
	Independent persons	Selbstandige Berufslose
	Dependent persons	Erhaltene Personen
B	Employment status	Stellung im Beruf
C	Industrial classification	Wirtschaftlicher Zugehörigkeit
	Division(s)	Wirtschaftsabteilungen
	Major group(s)	Betriebsklassen
	Group(s)	Betriebsgruppen
D	Occupational classification	Beruflicher Zugehörigkeit
	Major group(s)	Berufsabteilungen
	Minor group(s)	Berufsobergruppen
	Unit group(s)	Berufsgruppen
	Unit subgroup(s) (Individual occupations)	Berufsarten
	HOUSEHOLD AND FAMILY CHARACTERISTICS	
F	[Household]	...
	Private household	Privathaushalt
	One-person household	Einpersonenhaushalt
	Multi-person household	Mehrpersonenhaushalt
	Institutional household	Anstaltshaushalt

TABLE TITLES BY VOLUME

VOLUME 1. PRELIMINARY MAIN RESULTS OF THE POPULATION CENSUS OF MARCH 21, 1961, BY COMMUNE.

TABLES

Number[1]	Title	Page
I.	Principal summary.	7
II.	Federal States: Tabulations by Commune.	
	Wien.	19
	Niederösterreich.	20
	Burgenland.	55
	Oberösterreich.	63
	Salzburg.	75
	Steiermark.	79
	Kärnten.	98
	Tirol.	104
	Vorarlberg.	111

VOLUMES 2-11. BURGENLAND...AUSTRIA.

Structure of Volumes 2-11.

Data for the federal states of Austria are contained in Volumes 2-10, and Volume 11 contains data for Austria as a whole. Table numbers and table titles for all these volumes are identical. Page locations, however, vary from volume to volume. Hence, it is necessary to consult the Tables list as well as the Page Location Matrix below in order to locate desired data.

TABLES

Number	Title
1.	Principal results by commune.
2.	Residents by age group, marital status, sex, and political district.
3.	Residents by age, marital status, and sex.
4.	Austrians by age, marital status, and sex.

---

[1]Tables for the individual federal states are not numbered separately.

AUSTRIA

5. Residents by year of birth, marital status, and sex.
6. Austrians by year of birth, marital status, and sex.
7. Residents by major industrial group, sex, and political district.
8. Economically active population (employed and unemployed) and remaining income recipients by major industrial group, sex, and political district.
9. Residents by economic activity, industrial division, and sex.
10. Economically active population (employed and unemployed) and dependents by industrial group, and sex.
11. Residents by economic activity, age group, and sex.
12. Economically active population (employed and unemployed) by major industrial group, employment status, age group, and sex.
13. Economically active population (employed and unemployed) by individual occupation, sex, employment status, and age group.
14. Residents by type of household, age group, marital status, and sex.
15. Residents in private multi-person households by age group, position in household, and sex.
16. Residents in private and institutional households, by number of persons and political district.
17. Persons in one-person households by industrial division, employment status, and sex.
18. Private multi-person households by size of household, number of children under 14 years of age living in household, and industrial division of the head of the household.
19. Private multi-person households by size of household, number of children under 14 years of age living in household, and employment status of the head of the household.

TABLE TITLES BY VOLUME

PAGE LOCATION MATRIX

Table Number	Burgenland	Vorarlberg	Tirol	Salzburg	Kärnten	Niederösterreich	Oberösterreich	Steiermark	Wien	Österreich
1	8	8	8	8	8	8	8	8	8	8
2	18	12	18	14	16	58	24	34	10	9
3	20	13	20	15	18	62	27	37	14	11
4	22	15	22	17	20	64	29	39	16	13
5	24	17	24	19	22	66	31	41	18	15
6	26	19	26	21	24	68	33	43	20	17
7	28	21	28	23	26	70	35	45	22	19
8	30	22	30	24	28	74	36	46	24	20
9	32	22	32	24	30	78	38	48	28	22
10	33	24	33	26	31	79	40	50	29	24
11	44	25	44	27	42	90	41	51	40	25
12	45	36	45	38	43	91	52	62	41	36
13	56	47	56	49	54	102	63	73	52	47
14	67	58	67	60	65	113	74	84	63	58
15	68	59	68	61	66	114	75	85	64	59
16	68	59	68	61	66	115	75	85	64	59
17	69	60	69	62	67	115	76	86	65	60
18	69	60	69	62	67	116	76	86	65	60
19	69	60	69	62	67	116	76	86	65	60

AUSTRIA

VOLUME 12. HOUSEHOLDS IN AUSTRIA.

TABLES

Number	Title	Page
1.	Residents of private and institutional households by number of persons in political and juridical districts.	7
2.	Residents by type of household, age group, marital status, and sex.	13
3.	Residents in private multi-person households by age group, position in household, marital status, and sex.	14
4.	Persons in one-person households by industrial division, employment status, age group, and sex.	17
5.	Private multi-person households by size of household, number of children under 6 and under 14 years of age living in household, number of all dependents living in household, industrial division, and employment status of the head of the household.	18
6.	Private multi-person households by size of household, number of income recipients, economic activity, and employment status of the head of household.	20
7.	Private multi-person households by size of household, industrial division, age group, and sex of the head of household.	22
8.	Domestic servants living in private multi-person households by industrial division and employment status of employer.	24

VOLUME 13. COMPOSITION OF THE AUSTRIAN RESIDENT POPULATION BY GENERAL DEMOGRAPHIC AND CULTURAL CHARACTERISTICS.

TABLES

Number	Title	Page
1.	Increase of the resident population from 1951 to 1961 by sex, political district, and commune size-class.	6
2.	Communes and resident population by commune size-class and political district.	9
3.	Resident population by age, sex, and federal state.	14
4.	Austrians by age, sex, and federal state.	18
5.	Aliens by age, sex, and federal state.	22

6.	Resident population by year of birth, sex, and federal state.	26
7.	Austrians by year of birth, sex, and federal state.	30
8.	Aliens by year of birth, sex, and federal state.	34
9.	Resident population by age, marital status, and sex.	38
10.	Austrians by age, marital status, and sex.	40
11.	Aliens by age, marital status, and sex.	42
12.	Resident population by year of birth, marital status, and sex.	44
13.	Austrians by year of birth, marital status, and sex.	46
14.	Aliens by year of birth, marital status, and sex.	48
15.	Resident population by type of commune,[1] commune size-class, age group, sex, and federal state.	50
16.	Resident population by nationality, sex, and federal state.	68
17.	Resident population by religious affiliation, sex, and federal state.	70
18.	Resident population of Burgenland by usual language[2] and commune.	70
19.	Resident population of selected districts of Kärnten by usual language.[2]	76
20.	Resident population 14 years of age and older by highest-completed educational level,[3] age, sex, and federal state.	77
21.	Resident population 14 years of age and older by highest-completed educational level,[3] age, marital status, and sex.	80

---

[1] Communes are classified according to the number of persons engaged in agriculture in the commune. "Type of commune" refers to the degree of participation in agriculture.

[2] Usual language data are given only for districts in which in 1951 a number of persons with non-German usual language were recorded by the specific language.

[3] In determining the "highest-completed educational level," education is considered completed only if it has been terminated and a certificate received or the proper termination examinations taken. Thus anyone still in school beyond the secondary level (e.g., in the university) is considered to have completed only secondary schooling.

AUSTRIA

VOLUME 14. ECONOMICALLY ACTIVE POPULATION BY OCCUPATIONAL CLASSIFICATION.

TABLES

Number	Title	Page
1.	Economically active population (employed and unemployed) by individual occupation, employment status, age group, and sex.	9
2.	Economically active population (employed and unemployed) by occupational unit group, employment status, highest-completed educational level, and sex.	561
3.	Economically active alien population (employed and unemployed) by country of origin (OECD countries),[1] occupational minor group, employment status, and sex.	593
4.	Economically active female population (employed and unemployed) by occupational unit group, employment status, age group, and number of children under 14 years of age residing in household.	619

VOLUME 15. ECONOMICALLY ACTIVE POPULATION BY INDUSTRIAL CLASSIFICATION.

TABLES

Number	Title	Page
1.	Employed population by major industrial group, employment status, age group, and sex.	9
2.	Employed population by occupational unit group, major industrial group, and sex.	37
3.	Unemployed population by industrial group, employment status in last occupation, and sex.	109
4.	Unemployed population by major industrial group, employment status in last occupation, age group, and sex.	121
5.	Economically active alien population (employed and unemployed) by industrial group, employment status, and sex.	149
6.	Domestic servants living in private households by industrial division and employment status of their employers.	175

---

[1] OECD = Organization for European Cooperation and Development.

TABLE TITLES BY VOLUME

VOLUME 16. COMMUNES OF RESIDENCE--COMMUNES OF WORK OF EMPLOYED PERSONS IN AUSTRIA.[1]

TABLES[2]

Title	Page
Results by Commune:	13
Wien.	15
Niederösterreich.	31
Burgenland.	191
Oberösterreich.	225
Salzburg.	301
Steiermark.	323
Kärnten.	417
Tirol.	453
Vorarlberg.	491

VOLUME 17. RESIDENT POPULATION OF AUSTRIA BY SOURCE OF INCOME AND INDUSTRIAL CLASSIFICATION.

TABLES

Number	Title	Page
1.	Resident population by economic activity, major industrial group, sex, and political district.	8
2.	Resident population by economic activity, industrial division, commune size-class, and federal state.	74
3.	Resident population by economic activity, industrial division, average altitude of communes, and federal state.	79
4.	Resident population by economic activity, industrial division, employment status of earner, and federal state.	83

---

[1] This volume considers occupational commuting from one commune to another daily, weekly, monthly, or in some other interval. It excludes commuting by students and school children and intracommunal commuting. The patterns of worker movement and the working population (those employed and residing in the commune minus out-commuters, plus in-commuters) are ascertained.

[2] The tables in this volume are not numbered.

AUSTRIA

5. Resident population by economic activity, major industrial group, employment status of earner, and federal state. 88
6. Resident population by economic activity, age group, marital status, and sex. 108
7. Resident population by economic activity, major industrial group, employment status of earner, age group, and sex. 113
8. Resident population by economic activity, industrial division, nationality, and sex. 167
9. Resident population by economic activity, industrial group, nationality, and sex. 170

BELGIUM

MAIN ENTRY AND VOLUME NOTES:

Belgium. Institut National de Statistique.
    Recensement de la population, 31 décembre 1961.
Bruxelles, 1963-1966.
    10 v. 29cm.   (64-50426)

Tome 1. Chiffres de la population. 1963. 283 p. [Excluded; text tables only.]

Tome 2. Recensement des bâtiments. I. Royaume, provinces, et arrondissements. 1966. 149 p. II. Principaux résultats par commune. 1965. 116 p. [Excluded.]

Tome 3. Recensement des logements. I. Royaume, provinces, et arrondissements. 1965. 213 p. II. Principaux résultats par commune. 1965. 115 p. [Excluded.]

Tome 4. Répartition de la population selon l'état civil, la nationalité, et le lieu de naissance. I. Royaume, provinces, et arrondissements. 1966. 146 p. II. Principaux résultats par commune. 1965. 127 p.

Tome 5. Répartition de la population par âge. I. Royaume, provinces, et arrondissements. 1965. 301 p. II. Principaux résultats par commune. 1964. 111 p.

Tome 6. Recensement des ménages et des noyaux familiaux. I. Royaume, provinces, et arrondissements. 1966. 174 p. II. Principaux résultats par commune. 1965. 116 p.

Tome 7. Recensement des familles. 1966. 176 p. [Excluded; text tables only.]

Tome 8. Répartition de la population d'après l'activité, la profession et l'état social. I. Royaume, provinces, et arrondissements. 1966. 330 p. II. Principaux résultats par commune. 1965. 368 p.

Tome 9. Mobilité géographique de la main-d'oeuvre (Recensement des migrants alternants). 1965. 159 p.

Tome 10. Degré d'instruction de la population. I. Royaume, provinces, et arrondissements. 1966. 268 p. II. Principaux résultats par commune. 1965. 168 p.

BELGIUM

ENGLISH TRANSLATION OF CENSUS AND VOLUME TITLES:

Census of the population, December 31, 1961.

1. Population figures.
2. Census of buildings. Part I. Kingdom, provinces, and districts. Part II. Principal results by commune.
3. Census of dwellings. Part I. Kingdom, provinces, and districts. Part II. Principal results by commune.
4. Population distribution by marital status, nationality, and birthplace. Part I. Kingdom, provinces, and districts. Part II. Principal results by commune.[1]
5. Population distribution by age. Part I. Kingdom, provinces, and districts. Part II. Principal results by commune.
6. Census of households and family nuclei. Part I. Kingdom, provinces, and districts. Part II. Principal results by commune.
7. Census of families.
8. Population distribution by industry, occupation, and social status. Part I. Kingdom, provinces, and districts. Part II. Principal results by commune.
9. Geographical mobility of the labor force (Census of commuters).
10. Educational level of the population. Part I. Kingdom, provinces, and districts. Part II. Principal results by commune.

ENUMERATION DATE: December 31, 1961.

ENUMERATION BASIS: De jure.

GLOSSARY

ADMINISTRATIVE, GEOGRAPHICAL, AND POLITICAL UNITS

Level	English Translation	Vernacular Term
National	Kingdom of Belgium	Royaume de Belgique
Intermediate	Province	Province
	District	Arrondissement
Local	Commune	Commune

Provinces are the largest administrative divisions of Belgium. There are nine provinces, divided into a total of 26

---

[1] Part II does not include a tabulation by birthplace.

GLOSSARY

districts, which aid the provincial governments in carrying out their responsibilities.

Communes are local governmental units and number approximately 2,660.

TABLE OF DEMOGRAPHIC CONCEPTS

The demographic concepts which appear in the following table correspond to the recommendations of the European Conference of Statisticians, unless otherwise noted. They are defined and discussed in the appendixes and in footnotes. Terms in parentheses are alternative forms which appear in the table titles.

Appendix	English Table Term	Vernacular Table Term
	ECONOMIC CHARACTERISTICS	
A	Economic activity	Activité économique
	Economically active	Population active
	Economically inactive	Population non-active
B	Employment status	État socio-professionnel
C	Industrial classification	Activité économique
	Division(s)	Branches d'activités
	Major group(s)	Classes d'activités
	Group(s)	Groupes d'activités
D	Occupational classification[1]	Profession
	Major group(s) (Group or occupation)	Groupes de professions
	Minor group(s) (Group or occupation)	Groupes de professions
	Unit group(s) (Occupation)	Profession
E	Socio-economic category	Catégorie socio-professionnelle
	HOUSEHOLD AND FAMILY CHARACTERISTICS	
F	Household	Ménage
	Private household	Ménage ordinaire
	Institutional household	Ménage
	Family nucleus	Famille biologique

---

[1] In the table titles the term "occupation" usually designates a full tabulation of all three occupational groups (major, minor, and unit), but it may also designate only major and minor groups. The term "occupational groups" usually designates a tabulation of major and minor groups together, but it may also designate these groups individually.

BELGIUM

TABLE TITLES BY VOLUME

VOLUME 4. POPULATION DISTRIBUTION BY MARITAL STATUS, NATIONALITY, AND BIRTHPLACE. PART I. KINGDOM, PROVINCES, AND DISTRICTS.

TABLES

Number	Title	Page
1.	Distribution of the total population and the foreign population born in Belgium by sex, district of residence, and district of birth.	69
2.	Distribution of the total and foreign population, foreign born, by sex, district of residence, and country of birth.	89
3.	Distribution of foreign population by country of nationality, place of birth, sex, and district of residence.	105
4.	Distribution of foreigners by nationality and year of settlement in Belgium.	123
	Kingdom	124
	Provinces	126

VOLUME 4. POPULATION DISTRIBUTION BY MARITAL STATUS AND NATIONALITY. PART II. PRINCIPAL RESULTS BY COMMUNE.

TABLES

Number	Title	Page
1.	Population distribution of communes and administrative districts by sex and marital status.	
	Province of Anvers	10
	Province of Brabant	14
	Province of West Flandre	22
	Province of East Flandre	28
	Province of Hainaut	35
	Province of Liège	45
	Province of Limbourg	53
	Province of Luxembourg	58
	Province of Namur	63
	Summary table and results for the Kingdom	71
2.	Population distribution of communes and administrative districts by nationality.	
	Province of Anvers	74
	Province of Brabant	77
	Province of West Flandre	83

TABLE TITLES BY VOLUME

Province of East Flandre	88
Province of Hainaut	94
Province of Liège	102
Province of Limbourg	109
Province of Luxembourg	113
Province of Namur	117
Summary table and results for the Kingdom	123

VOLUME 5. POPULATION DISTRIBUTION BY AGE. PART I. KINGDOM, PROVINCES, AND DISTRICTS.

TABLES

Number	Title	Page
1.	Distribution of the total population and foreigners by year of birth, sex, and marital status.	
	Kingdom	82
	Provinces	86
	Administrative districts	122
2.	Distribution of foreigners by age group,[1] sex, and nationality.	
	Kingdom	288
	Provinces	288

VOLUME 5. POPULATION DISTRIBUTION BY AGE. PART II. PRINCIPAL RESULTS BY COMMUNE.

TABLES

Number	Title	Page
1.	Population distribution of communes, administrative districts, and provinces by age group[1] and sex.	
	Province of Anvers	8
	Province of Brabant	14
	Province of West Flandre	26
	Province of East Flandre	36
	Province of Hainaut	48
	Province of Liège	64
	Province of Limbourg	78
	Province of Luxembourg	86
	Province of Namur	94
	Summary table and results for the Kingdom	106

---

[1] Five-year age groups.

BELGIUM

VOLUME 6. CENSUS OF HOUSEHOLDS AND FAMILY NUCLEI. PART I. KINGDOM, PROVINCES, AND DISTRICTS.[1]

TABLES

Number	Title	Page
1.	Distribution of private households by type of household, sex of the head of the household, and composition.	
	Kingdom	52
	Provinces	53
2.	Distribution of private households by type of household and by the age, sex, and marital status of the head of the household.	64
3.	Distribution of private households by the sex, marital status, and age group of the head of the household and composition of the household.	
	Kingdom	72
	Provinces	76
4.	Distribution of private households by the occupation of the head of the household and the composition of the household.	
	Kingdom	114
	Provinces	116
5.	Distribution of private households by the occupational group, marital status, age group, and sex of the head of the household.	136
6.	Distribution of private households by province and by the nationality and sex of the head of the household.	144
7.	Distribution of private households by the nationality and sex of the head of the household and the composition of the household.	
	Absolute figures	148
	Relative figures	150
8.	Distribution of institutional households by type of household and composition.	
	Kingdom	154
	Provinces	154

---

[1] Volume 6, Parts I and II are based on the household census, and thus do not correspond to the de jure enumeration of the total population. In the household census, some households that were temporarily absent at the time of the census, but have their usual residence in Belgium, were not included.

TABLE TITLES BY VOLUME

9. Distribution of family nuclei classified by composition and type and by the type of household.
    Kingdom 158
    Provinces 158

10. Distribution of family nuclei classified by type and by the number of unmarried children.
    Kingdom 170
    Provinces 170

VOLUME 6. CENSUS OF HOUSEHOLDS AND FAMILY NUCLEI. PART II. PRINCIPAL RESULTS BY COMMUNE.[1]

TABLES

Number	Title	Page
1.	Number of households by type and composition.	
	Province of Anvers	12
	Province of Brabant	15
	Province of West Flandre	21
	Province of East Flandre	26
	Province of Hainaut	32
	Province of Liège	40
	Province of Limbourg	47
	Province of Luxembourg	51
	Province of Namur	55
	Summary table and results for the Kingdom	61
2.	Number of family nuclei by type and number of members.	
	Province of Anvers	64
	Province of Brabant	67
	Province of West Flandre	73
	Province of East Flandre	78
	Province of Hainaut	84
	Province of Liège	92
	Province of Limbourg	99
	Province of Luxembourg	103
	Province of Namur	107
	Summary table and results for the Kingdom	113

---

[1] See footnote to Volume 6, Part I.

BELGIUM

VOLUME 8. POPULATION DISTRIBUTION BY INDUSTRY, OCCUPATION, AND SOCIAL STATUS. PART I. KINGDOM, PROVINCES, AND DISTRICTS.[1]

TABLES

Number	Title	Page
1.	Distribution of the economically active population by industrial group, sex, and employment status. Total population and foreigners.	
	Kingdom	129
2.	Distribution of the economically active population by industrial major group, sex, and employment status. Total and foreign populations.	
	Kingdom	150
	Provinces	154
3.	Distribution of the economically active population by industrial major group, age group, sex, and employment status.	
	Kingdom	191
4.	Distribution of the economically active population (unemployed and militiamen excluded) by industrial major group, marital status, sex, and employment status.	
	Kingdom	207

---

[1] This volume is based on the general population census, but a census of industry and commerce was also taken in 1961. These two censuses are not comparable for the following reasons: (1) In the general population census, each economically active person was asked to give the nature of the industry in which he was occupied, which was done in very general terms and only for the main industry of the establishment. In the industry and commerce census, the information was given by the head of the establishment and was much more detailed. Therefore the distribution is not identical, but more precise in the industry and commerce census. (2) In the general population census, everyone whose usual residence was in Belgium was included. In the industry and commerce census, only persons employed in Belgian enterprises were included. The 75,000 Belgians who work in foreign countries were thus included in the former census, but omitted in the latter. (3) The industry and commerce census did not include all part-time employees, while the general population census did. (4) In the general population census some blue collar workers (ouvriers) claimed to be white collar workers (employés), whereas in the census of industry and commerce, the classification was done by the employers themselves according to the classification system for that census.

5.	Distribution of the economically active population (unemployed and militiamen excluded) by occupation, sex, and employment status. Kingdom	223
5b.	Distribution of the economically active foreign population (unemployed and militiamen excluded) by occupational group, sex, and employment status. Kingdom	245
6.	Distribution of the economically active population (unemployed and militiamen excluded) by occupational group, marital status, sex, and employment status. Kingdom	251
7.	Distribution of retired and pensioned persons by their prior occupation, sex, and employment status. Kingdom	273
8.	Distribution of the unemployed by last occupation, sex, and employment status. Kingdom	279
9.	Distribution of militiamen by last occupation and employment status. Kingdom	285
10.	Distribution of the economically active population by industrial major group, occupation, sex, and employment status. Kingdom	289
11.	Distribution of the economically active population by secondary occupation, sex, and employment status. Kingdom	315
12.	Distribution of the economically active population by secondary occupation, primary occupation, and sex. Kingdom	323

VOLUME 8. POPULATION DISTRIBUTION BY INDUSTRY, OCCUPATION, AND SOCIAL STATUS. PART II. PRINCIPAL RESULTS BY COMMUNE.

TABLES

Number	Title	Page
1.	Population distribution by employment status and sex. Results by commune.	
	Province of Anvers	12
	Province of Brabant	18
	Province of West Flandre	30
	Province of East Flandre	40
	Province of Hainaut	52
	Province of Liège	68

BELGIUM

        Province of Limbourg   82
        Province of Luxembourg   90
        Province of Namur   98

        Summary table and results for the Kingdom   110

2. Distribution of the economically active population (unemployed and militiamen excluded) by commune of residence, industrial division, and sex.
        Province of Anvers   114
        Province of Brabant   122
        Province of West Flandre   138
        Province of East Flandre   150
        Province of Hainaut   164
        Province of Liège   184
        Province of Limbourg   200
        Province of Luxembourg   210
        Province of Namur   220

        Summary table and results for the Kingdom   236

3. Distribution of the economically active population (unemployed and militiamen excluded) by commune of work in Belgium, industrial division, and sex.
        Province of Anvers   240
        Province of Brabant   248
        Province of West Flandre   262
        Province of East Flandre   274
        Province of Hainaut   288
        Province of Liège   308
        Province of Limbourg   324
        Province of Luxembourg   334
        Province of Namur   344

        Summary table and results for the Kingdom   360

4. Distribution of the economically active population (unemployed and militiamen excluded) working in a foreign country by the country where employed, industrial division, and sex.   363

VOLUME 9. GEOGRAPHICAL MOBILITY OF THE LABOR FORCE (CENSUS OF COMMUTERS)[1]

TABLES

Number	Title	Page
1.	Distribution, on December 31, 1961, of outgoing commuters by means of transportation used and length of journey. Results by commune.	
	Province of Anvers	24

---

[1] A commuter is defined as a worker who works outside his commune

Province of Brabant	27
Province of West Flandre	33
Province of East Flandre	38
Province of Hainaut	44
Province of Liège	52
Province of Limbourg	59
Province of Luxembourg	63
Province of Namur	67
Summary table and results for the Kingdom	74

2. Distribution, on December 31, 1961, of incoming commuters by means of transportation used and length of journey. Results for the communes where more than 150 commuters work. Comparison with 1947.

Province of Anvers	76
Province of Brabant	78
Province of West Flandre	80
Province of East Flandre	82
Province of Hainaut	84
Province of Liège	87
Province of Limbourg	89
Province of Luxembourg	90
Province of Namur	91

3. Geographical mobility of the labor force (unemployed and militiamen excluded) and employment coefficient by commune.

Province of Anvers	94
Province of Brabant	97
Province of West Flandre	104
Province of East Flandre	109

---

of residence. The following questions were asked of every economically active person: What is the address of your place of work? If the place where you work is different from where you live: (1) Do you commute five days a week between your dwelling and your place of work? (2) If yes: What means of transportation do you use? How much time does your commuting take? (3) If no: Do you have, during the week, a dwelling closer to your place of work than your residence? Both outgoing commuting and incoming commuting were ascertained. With respect to a given commune, outgoing commuters are those people who reside there but work in another commune. Incoming commuters are those people who work there but reside in another commune. The data covered every commune separately except the 19 communes of Brussels, which were taken together--consequently, workers from one commune in Brussels who work in another commune in the city were not counted as commuters. The employment coefficient in Table 3 is the ratio of the economically active population living in the commune to the economically active population working in the commune.

BELGIUM

Province of Hainaut	115
Province of Liège	124
Province of Limbourg	131
Province of Luxembourg	135
Province of Namur	140

4. Distribution of commuters of the Brussels population aggregate by commune of usual residence. Summary of communes containing more than 25 commuters.   148

VOLUME 10. EDUCATIONAL LEVEL OF THE POPULATION. PART I. KINGDOM, PROVINCES, AND DISTRICTS.[1]

TABLES

Number	Title	Page
1.	Distribution of the school population 11 years old and over by sex, age, and kind of education.	
	Kingdom	66
	Provinces	67
2.	Distribution of persons 14 years old and over who are no longer full-time students by sex, age at which full-time education was terminated, and kind of education pursued.	
	Kingdom	78
	Provinces	79
	Administrative districts	88
3.	Distribution of persons 14 years old and over who are no longer full-time students by sex, age group, age at which full-time education was terminated, and kind of education pursued.	
	Kingdom	130
4.	Distribution of persons 14 years old and over who	

---

[1] The data in Volume 10 relate only to full-time education. Full-time education is defined as daily instruction which is usually given during the day and for an entire academic year. It excludes part-time courses, evening courses, Sunday courses, correspondence courses, courses accompanying an apprenticeship in a firm, etc. All persons 11 years old and over were asked the following questions: (1) Are you still pursuing a full-time education? (2) If yes: (a) What kind of studies do you pursue? (b) In what commune is the school you attend found? (3) If no: (a) At what age did you quit pursuing a full-time education? (b) In what type of studies did you complete your last year of school? They were also asked if they had one of a number of diplomas listed on the form. If they had more than one diploma, they were classified by the most advanced.

	are no longer full-time students by occupational group, sex, and age at which full-time education was terminated.	
	Kingdom	136
5.	Distribution of the enumerated population with a certificate or diploma from a secondary, or higher, educational institution by sex and kind of certificate or diploma (highest level).	
	Kingdom	150
	Provinces	150
6.	Distribution of the enumerated population with a certificate or diploma from a secondary, or higher, educational institution by economic activity, employment status, sex, and kind of certificate or diploma (highest level).	
	Kingdom	160
7.	Distribution of the population (unemployed and militiamen excluded) with a certificate or diploma from a secondary, or higher, educational institution by age group, occupational group, sex, and kind of certificate or diploma (highest level).	
	Kingdom	172
8.	Distribution of the economically active population (unemployed and militiamen excluded) with a certificate or diploma from a secondary, or higher, educational institution (highest level) by sex and industrial major group.	
	Kingdom	244

VOLUME 10. EDUCATIONAL LEVEL OF THE POPULATION. PART II. PRINCIPAL RESULTS BY COMMUNE.[1]

TABLES

Number	Title	Page
1.	Distribution of the population 11 years old and over who are still full-time students by the kind of education pursued.	
	Province of Anvers	12
	Province of Brabant	15
	Province of West Flandre	21
	Province of East Flandre	26
	Province of Hainaut	32
	Province of Liège	40
	Province of Limbourg	47

---

[1] See footnote for Volume 8, Part I.

BELGIUM

	Province of Luxembourg	51
	Province of Namur	55
	Summary table and results for the Kingdom.	62
2.	Distribution of persons 14 years old and over who are no longer receiving instruction by kind of education pursued and certificate, diploma, or certificate of completion of studies (highest level).	
	Province of Anvers	64
	Province of Brabant	70
	Province of West Flandre	82
	Province of East Flandre	92
	Province of Hainaut	104
	Province of Liège	120
	Province of Limbourg	134
	Province of Luxembourg	142
	Province of Namur	150
	Summary table and results for the Kingdom	164

DENMARK

MAIN ENTRIES AND VOLUME NOTES:

Denmark. Statistiske Departement.
Folke- og boligtaellingen 26. september 1960. København,
1963-1965.
3 v. 29cm. (NUC65-94782)

A. Folke- og boligtal i kommuner og bymaessige bebyggelser 1960. (1) Øerne undtagen hovedstaden. 1963. 165 p. (2) Jylland. 1963. 219 p. (3) Hovedstaden med forstaeder. Sammendrag. 1963. 95 p.

B. Fordelinger efter alder og erhverv m.v. (1) Køn, alder og aegteskabelig stilling. 1963. 95 p. (2) Erhverv og arbejdsstilling. 1964. 172 p. (3) Fag, arbejdsstilling og erhverv. 1964. 151 p.

C. [1] Bolig- og husstandsundersøgelse 1960. 1964. 159 p. [Excluded.] [2] Boligforholdene i byer og bymaessige bebyggelser med over 1,000 indbyggere 1960. 1964. 197 p. [Excluded.]

Denmark. Statistiske Departement.
Faerøerne. Folketaellingen 1960. Aegteskaber, fødte og døde, 1946-60. København, 1965.
44 p. 29cm.

Denmark. Statistiske Departement.
Grønland. Folketaellingen 1960. Aegteskaber, fødte og døde, 1952-60. København, 1965.
85 p. 29cm.

ENGLISH TRANSLATION OF CENSUS AND VOLUME TITLES:

Denmark.

Population and housing census, September 26, 1960.

A. Population and housing in communes and densely populated areas.
   (1) The islands excluding the capital cities.
   (2) Jutland.
   (3) Capital cities with suburbs. Summary.
B. Distribution by age, industry, etc.
   (1) Sex, age, and marital status.
   (2) Industry and employment status.
   (3) Occupation, employment status, and industry.
C. [Housing and households]
   [1] Housing and households. 1960.

DENMARK

[2] Housing conditions in towns and urban areas with over 1,000 inhabitants in 1960.

Faroe Islands.

Population census 1960. Marital status, births, and deaths, 1946-1960.

Greenland.

Population census 1960. Marital status, births, and deaths, 1952-1960.

ENUMERATION DATE: September 26, 1970.

ENUMERATION BASIS: De jure.

GLOSSARY

ADMINISTRATIVE, GEOGRAPHICAL, AND POLITICAL UNITS

Level	English Translation	Vernacular Term
National	Kingdom of Denmark	Kongeriget Danmark
Intermediate	County or County council district	Amt or Amsrådskreds
Local	Commune	Kommune
	Rural commune	Sognekommune
	Urban commune	Bykommune
Population cluster ("locality") concept:[1]		
	Densely populated area	Bymaessig bebyggelse
Other concepts:		
	Capital city district	Hovedstadsområde
	Capital city	Hovedstadt
	Suburb	Forstadskommune
	Surrounding commune	Omegnskommune
	Principal parts of the country	Landsdele and hovedlandsdele
	Provincial town	Provinsbyer

Denmark is composed of 22 <u>counties</u> for administrative purposes.[2] Three of the counties are divided into two council districts each, which, taken together with the 19 council districts formed by the remaining undivided counties, constitute the nation's 25 <u>county council districts</u>.

---

[1] For a discussion of this concept, see European Population Censuses: The 1960 Series, U.N. Doc.: ST/CES/3 (1964), p. 108.

[2] Statesman's Yearbook 1967-1968 (New York: St. Martin's Press, 1967), p. 945.

34

GLOSSARY

The county council districts are composed of two kinds of communes, which are the units of local government in Denmark: rural communes and urban communes.

Rural communes vary greatly in size and population, and may surpass certain urban communes in number of inhabitants. Urban communes are independent of the districts in which they are situated,[1] and include Copenhagen, Frederiksberg, and Gentofte, along with the provincial towns.

Densely populated areas are rural communes with a population of 200 or more, and in the census they are listed alphabetically for each county. When such areas lie inside more than one commune, figures for the total population of the area, as well as for the component parts, are provided. Densely populated areas situated in two counties are listed in both counties.

The capital city district consists of the capital city, Copenhagen; the two cities contiguous to it, Frederiksberg and Gentofte; its suburbs; and the surrounding communes.

Principal parts of the country are geographical parts of the country (such as the islands or Jutland) and political parts (such as the capital city or rural communes).

Provincial towns are towns with suburbs of more than 700 persons.

TABLE OF DEMOGRAPHIC CONCEPTS

The demographic concepts which appear in the following table correspond to the recommendations of the European Conference of Statisticians, unless otherwise noted. They are defined and discussed in the appendixes and in footnotes. Terms in parentheses are alternative forms which appear in the table titles. Terms in brackets are not used in the table titles, but they are included here in order to give logical completeness to the various classifications in the table.

Appendix	English Table Term	Vernacular Table Term
	ECONOMIC CHARACTERISTICS	
A	[Economic activity]	...
	Economically active	De i erhvervene beskaeftigede personer

---

[1] Samuel Humes and Eileen M. Martin, The Structure of Local Governments Throughout the World (The Hague: Martinus Nijhoff, 1961), p. 265.

DENMARK

B	Employment status	Arbejdsstilling
C	Industrial classification 　Division(s) 　Major group(s) 　Group(s)	Erhverv Erhvervshovedgrupper Erhvervsgrupper Enkelte erhverv
D	Occupational classification[1] 　Major group(s) (Occupation) 　Minor group(s) (Occupation) 　Unit group(s) (Occupation)	Fag Fag Fag Fag
E	Socio-economic category	Socioøkonomisk gruppe

HOUSEHOLD AND FAMILY CHARACTERISTICS

F	[Household] 　Private household 　Institutional household Marital status[2]	... Privat husstand Hushold Aegteskabelig

TABLE TITLES BY VOLUME

A. POPULATION AND HOUSING IN COMMUNES AND DENSELY POPULATED AREAS. (1) THE ISLANDS EXCLUDING THE CAPITAL CITIES.

TABLES[3]

Number	Title
1.	Total population by (a) age and sex and (b) industrial division.
2.	Total population by employment status and industrial division.
3.	Dwellings by kind of property and number of rooms.
4.	Private households by industry[4] of the head of the household; institutional households, etc.

---

[1] In the table titles, "occupation" designates throughout tabulations of one, two, or all three occupational groups.

[2] Marital status comprises the following categories: unmarried, married, widowed, divorced, and separated.

[3] No page numbers given.

[4] Agricultural or other industry.

TABLE TITLES BY VOLUME

A. POPULATION AND HOUSING IN COMMUNES AND DENSELY POPULATED AREAS. (2) JUTLAND.

TABLES[1]

Number	Title
1.	Total population by (a) age and sex and (b) industrial division.
2.	Total population by employment status and industrial division.
3.	Dwellings by kind of property and number of rooms.
4.	Private households by industry[2] of the head of the household; institutional households, etc.

A. POPULATION AND HOUSING IN COMMUNES AND DENSELY POPULATED AREAS. (3) CAPITAL CITIES WITH SUBURBS. SUMMARY.

TABLES[1]

Number — Title

Capital city district

1. Total population by (a) age and sex and (b) industrial division.
2. Total population by employment status and industrial division.
3. Dwellings by kind of property and number of rooms.

<u>Total country: Distribution by county and county council district, etc., by principal parts of the country</u>

1. Total population by (a) age and sex and (b) industrial division.
2. Total population by employment status and industrial division.
3. Dwellings by kind of property and number of rooms.
4. Private households by industry[2] of the head of the household; institutional households, etc.

---

[1] No page numbers given.

[2] Agricultural or other industry.

DENMARK

Total country: Distribution by trade areas and trade districts[3]

1. Total population by (a) age and sex and (b) industrial division.
2. Total population by employment status and industrial division.
3. Dwellings by kind of property and number of rooms.
4. Private households by industry[2] of head of household; institutional households, etc.

B. DISTRIBUTION BY AGE, INDUSTRY, ETC. (1) SEX, AGE, AND MARITAL STATUS.

TABLES

Number	Title	Page
1.	Population by sex, single years of age, and marital status. Principal parts of the country.	6
2.	Population by sex, single years of age, and marital status. Towns with over 30,000 inhabitants.	26
3.	Population by sex, 5-year age groups, and marital status. Capital city district and provincial towns.	50
4.	Population by sex, 5-year age groups, and marital status. Counties and county council districts.	63
5.	Population by sex, year of birth, and marital status. Principal parts of the country.	76
6.	Economically active persons and domestic help by principal parts of the country, sex, and single years of age.	90
7.	Population by sex, age, and marital status. Principal parts of the country. 1950 and 1960.	94

B. DISTRIBUTION BY AGE, INDUSTRY, ETC. (2) INDUSTRY AND EMPLOYMENT STATUS.

TABLES

Number	Title	Page
1.	Population by industrial division. Principal parts of the country. 1960.	6

[3]There are five trade areas (handelsområder) which are composed of 44 trade districts (handeldistrikter).

38

TABLE TITLES BY VOLUME

2a.	Population by major industrial group. Principal parts of the country. 1960.	10
2b.	Population by industrial group. Principal parts of the country. 1960.	22
3.	Economically active persons, rentiers, etc. Capital city district. 1960.	72
4.	Economically active persons, rentiers, etc. Provincial towns. 1960.	88
5.	Industrial major group and employment status. Counties, county council districts. 1960.	136
6.	Census population by industrial group and employment status. 1950 and 1960.	162

B. DISTRIBUTION BY AGE, INDUSTRY, ETC. (3) OCCUPATION, EMPLOYMENT STATUS, AND INDUSTRY.

TABLES

Number	Title	Page
1.	Population by age and industrial major group. Whole country. 1960.	6
2.	Economically active persons by employment status, sex, age, and major industrial group. Whole country. 1960.	6
3.	Economically active persons by employment status, region, and major industrial group. 1960.	11
4.	Population by occupation and (a) age, (b) sex, and (c) region. 1960.	16
5a.	Economically active persons in counties and towns with over 7,000 inhabitants by employment status and occupation. 1960.	60
5b.	Housewives remaining at home by age and husband's industry. Children, rentiers, etc. by age. Counties and towns with over 7,000 inhabitants. 1960.	68
6.	Public servants and workers by occupation and industrial major group. Whole country. 1960.	74
7.	Economically active persons by major industrial group and occupation. Whole country. 1960.	84
8.	Population in main regions by socio-economic category and position in household. 1960.	102
9a.	Economically active persons by socio-economic	

DENMARK

	category, employment status, sex, age, and principal parts of the country. 1960.	106
9b.	Rentiers, dependents, etc. by socio-economic category, sex, age, and region. 1960.	116
10.	Total population and economically active persons by socio-economic category. Capital city district. 1960.	118
11.	Total population and economically active persons by socio-economic category. Provincial towns. 1960.	122
12.	Total population and economically active persons by socio-economic category. Counties, county council districts. 1960.	134

FAROE ISLANDS. POPULATION CENSUS 1960. MARITAL STATUS, BIRTHS, AND DEATHS, 1946-60.

TABLES

Number	Title	Page
1.	Islands and districts.	7
2.	Alphabetical list of communes.	8
3.	Parishes.	9
4.	Public health districts.	9
5.	Number of persons and households in the districts, communes, and towns. Population census September 26, 1960.	10
6.	Total population by (a) age and sex, (b) industrial division. 1960.	12
7.	Total population by employment status and abbreviated industrial division. 1960.	16
8.	Private households by abbreviated industrial division of the head of the household. 1960.	20
9.	Population by employment status, sex, and industrial division. Principal parts of the country. 1960.	22
10.	Population by employment status, sex, and industrial major group. 1960.	24
11.	Economically active persons by employment status and total population by industrial group. 1960.	26
12.	Population by sex, 5-year age groups, and marital status. September 26, 1960.	29

13.	Population by sex and single years of age. September 26, 1960.	30
14.	Population by sex and year of birth. September 26, 1960.	31
15.	Population by occupation and age. 1960.	32
16.	Population by socio-economic category. Principal parts of the country. 1960.	34
17.	Population by socio-economic category and position in the household. 1960.	34
18.	Economically active persons, pensioners, etc. by sex, age, and socio-economic category. 1960.	35
19.	Number of live births, deaths, etc., and marriages per year. 1921-1962.	37
20.	Number of births, deaths, and marriages, 1946-1960, by month.	38
21.	Number of births and number of children born, 1946-1960, by mother's age at birth of child.	39
22.	Percentage distribution of births, live and still, by mother's age at birth of child.	40
23.	Number of deaths, 1946-1960, by sex, age, and marital status.	40
24.	Death rate per 1000 in various age groupings. 1946-1960.	42
25.	Infant death rate (number of deaths under age one year per 1000 live births). 1946-60.	42
26.	Number of deaths under age one year. 1956-1960.	42
27.	Number of marriages, 1946-1960, by age and marital status of bride and bridegroom.	43
28.	Number of persons marrying per 1000 unmarried persons in the 15 to 59 year age class.	44
29.	Percentage distribution of the marital status of marrying persons before marriage.	44

GREENLAND. POPULATION CENSUS 1960. MARITAL STATUS, BIRTHS, AND DEATHS, 1952-60.

TABLES

Number	Title	Page
1.	Communes (districts) and jurisdictions. 1960.	7
2.	Provincial council districts. 1960.	7

DENMARK

3.	Police districts. 1960.	7
4.	Parishes. 1960.	7
5.	Electoral districts of the general election. 1960.	8
6.	Public health districts. 1960.	8
7.	School districts. 1960.	8
8.	Trade districts. 1960.	8
9a.	Administrative divisions. 1960. Register of communes.	9
9b.	Administrative divisions. 1960. Alphabetical register of towns, trading posts, settlements, etc.	10
10a.	Population by commune and residence district. Total population. 1960.	20
10b.	Population by commune and residence district. Persons born in Greenland. 1960.	30
11a.	Population by sex, age, and marital status. Total population. 1960.	40
11b.	Population by sex, age, and marital status. Persons born in Greenland. 1960.	41
12a.	Population in individual regions of the country by sex and age. Total population. 1960.	42
12b.	Population in individual regions of the country by sex and age. Persons born in Greenland. 1960.	43
13.	Population by sex, age, and place of birth of the respondent and his parents.	44
14a.	Population by industrial division, employment status, and sex. Total population. 1960.	46
14b.	Population by industrial division, employment status, and sex. Persons born in Greenland. 1960.	47
15.	Economically active persons, pensioners, etc. by industrial major group. 1960.	48
16.	Economically active persons by industrial major group, occupational minor group, and sex. 1960.	52
17.	Population by employment status, public or private sector, sex, and age. 1960.	56
18.	Population by respondent's or (economic) supporter's source of income, employment status, public or private sector. 1960.	58
19.	Economically active persons by industrial division, sex, source of income, employment status, public or private sector. 1960.	60

TABLE TITLES BY VOLUME

20.	All persons by relationship to the head of the household, sex, and age. 1960.	64
21.	All persons in private households by relationship to the head of the household, sex, and age. 1960.	66
22.	All persons in private households by family composition,[1] relation to the head of the household, sex, and age. 1960.	68
23.	Size of private households by region and commune. 1960.	70
24.	Size of private households by number of children in the household. 1960.	70
25.	Private households by number of families and children in the household. 1960.	72
26.	Families by size and number of children in the family. 1960.	72
27.	Families by number of persons and number of children in the family, by the marital status of the head of the family. 1960.	73
28.	Families by number of persons and number of children in the family, by the birthplace of the head of the family. 1960.	74
29.	Number of births, deaths, and marriages by individual year.	75
30.	Number of births, deaths, and marriages, 1952-1960, by month.	76
31.	Live and still births 1952-1955 and 1956-1960 by mother's age at the time of birth.	76
32.	Number of deaths 1952-1960 by sex, age, and marital status.	77
33.	Number of marriages 1952-1955 and 1956-1960 by bride's and bridegroom's ages and marital status.	78

---

[1] See Appendix F: "Family Composition of Private Households."

FINLAND

MAIN ENTRY AND VOLUME NOTES:[1]

Finland. Tilastollinen Päätoimisto.[2]
 Yleinen väestölaskenta. Allmänna folkräkningen. General census of population. 1960. Helsinki, 1962-1965.
 13 v. 30cm.

[Osa] I. Asuntokanta. Bostadsbeståndet. Housing. 1962. 164 p. [Excluded.]

[Osa] II. Väestön ikä sivilisääty, pääkieli ym. Befolkningens ålder, civilstånd, huvudspråk mm. Population by age, marital status, main language, etc. 1963. 134 p.

[Osa] III. Ammatissa toimivan väestön elinkeino ja ammattiasema. Förvärvsarbetande befolkning efter näringsgren och yrkesställning. Economically active population by industry and industry status. 1963. 73 p.

[Osa] IV. Väestön elinkeino ja ammattiasema. Befolkningen efter näringsgren och yrkesställning. Population by industry and industrial status. 1963. 191 p.

[Osa] V. Perheet. Familjer. Families. 1964. 81 p.

[Osa] VI. Väestön sosio-ekonominen asema; ammatissa toimivan väestön ikä, työpaikan sijainti. Befolkningen efter socio-ekonomisk ställning; den förvärvsarbetande befolkningen efter ålder, arbetsplatsens belägenhet. Population by socio-economic status; economically active population by age, location of working place. 1963. 77 p.

[Osa] VII. Ruokakunnat ja niiden asuminen. Hushållen och deras Bostadsförhållanden. Households and their housing conditions. 1963. 148 p.

---

[1] The census of Finland is published in three languages: Finnish, Swedish, and English. The main entry and volume notes are given here in all three languages; the glossary, in Finnish and English only; and the table titles, in English only.

[2] The National Union Catalog (Card No. NUC64-30111) uses the Swedish name for the Finnish statistical office: "Statistiska Centralbyrån." The Library of Congress, however, prefers the Finnish terminology as it appears here.

GLOSSARY

[Osa] VIII. Syntymäpaikka, koulusivistys, siirtoväki ym. Födelseort, skolbildning, förflyttad befolkning mm. Population by birth-place and by education, displaced population, etc. 1964. 63 p.

[Osa] IX. Ammatti ja ammattikoulutus. Yrke och yrkesutbildning. Occupation and vocational training. 1964. 155 p.

[Osa] X. Rakennuskanta. Byggnadsbeståndet. Buildings. 1964. 320 p. [Excluded.]

[Osa] XI. Taajamat ja niiden rajat, ym. Tätorterna och deras gränser, etc. Non-administrative urban settlements and their boundaries, etc. 1965. 127 p.

[Osa] XII. Täydennysosa. Kompletteringsvolym. Supplementary volume. 1965. 127 p.

[Osa] XIII. Taululuettelot. Tabellförteckningar. Lists of tables. 1965. 19 p. [Excluded.]

ENUMERATION DATE: December 31, 1960.

ENUMERATION BASIS: De jure.

GLOSSARY

ADMINISTRATIVE, GEOGRAPHICAL, AND POLITICAL UNITS

Level	English Translation	Vernacular (Finnish) Term
National	Republic of Finland	Suomen Tasavalta
Intermediate	Province	Lääni
Local	Commune	Kunta
	Rural communes	Maalaiskunnat
	Urban communes	Kaupungit ja kauppalat

Population cluster ("locality") concept:[1]

	Non-administrative urban settlement	Taajama

Other concept:

	Statistical region	Tilastoalue

For purposes of governmental administration, Finland is divided into 12 **provinces**. The provinces are governed by governors

---

[1] For a discussion of this concept see European Population Censuses: The 1960 Series, U.N. Doc.: ST/CES/3 (1964), p. 108.

FINLAND

appointed by the president, and are in turn divided into <u>communes</u>, the units of local government. There are two kinds of communes: <u>urban communes</u> (cities and towns) and <u>rural communes</u>. Urban and rural communes have almost the same structure and similar functions, but the latter receive larger subsidies from the central government.

<u>Non-administrative urban settlements</u> are groups of buildings, each building occupied by at least 200 people and not farther than 200 meters apart. They may include parts of one or more administrative communes.

<u>Statistical regions</u> are the administrative units which prepare annual registration lists.

TABLE OF DEMOGRAPHIC CONCEPTS

The demographic concepts which appear in the following table correspond to the recommendations of the European Conference of Statisticians, unless otherwise noted. They are defined and discussed in the appendixes and in footnotes. Terms in parentheses are alternative forms which appear in the table titles. Terms in brackets are not used in the table titles, but they are included here in order to give logical completeness to the various classifications in the table.

Appendix	English Table Term	Vernacular (Finnish) Table Term
CULTURAL, EDUCATIONAL, AND PERSONAL CHARACTERISTICS		
	Congregation[1]	Uskontokunta
ECONOMIC CHARACTERISTICS		

A	Economic activity
	   Economically active[2]
	   [Economically inactive]
		Ammattitoiminta
		   Ammatissa toimiva väestö
		   ...

B	Employment status (Industrial status)	Ammattiasema
C	Industrial classification Division(s) (Industry 1-digit level)	Elinkeino Elinkeinon (1-numeroinen rhymittely)

---

[1] Designates tabulations by religious denomination as follows: Lutheran, Orthodox Christian, Finnish Catholic, other Christian, non-Christian, Finnish civil register, unclear, and unknown.

[2] Persons, including family workers, 14 years of age or older, who work at least half-time. No distinction is made between employed and unemployed.

TABLE TITLES BY VOLUME

	Major group(s) (Industry 2-digit level)	Elinkeinon (2-numeroinen rhymittely)
	Group(s) (Industry 3-digit level)	Elinkeinon (3-numeroinen rhymittely)
D	Occupational classification[3]	Ammatti
	Major group(s) (Occupation)	Ammatti
	Minor group(s) (Occupation)	Ammatti
	Unit group(s) (Occupation)	Ammatti
E	Socio-economic category (Socio-economic status)	Sosio-ekonominen asema

HOUSEHOLD AND FAMILY CHARACTERISTICS

F	Households[4]	Ruokakuntia
	Family	Perhe

TABLE TITLES BY VOLUME

II. POPULATION BY AGE, MARITAL STATUS, MAIN LANGUAGE, ETC.

TABLES

Number	Title	Page
1.	Population and some important data on the structure of population, by provinces, statistical regions, communes and non-administrative urban settlements.	2
2.	Population by age, sex and marital status, by single years, in whole country, urban and rural communes and in non-administrative urban settlement areas in rural communes.	88

---

[3] In the table titles, "occupation" designates throughout tabulations of one, two, or all three occupational groups.

[4] In the table titles, the term "households" refers to private households only. These are classified into three groups:
- A. **Occupant:** Those households where a member of the household is the proprietor or principal tenant.
- B. **Sub-tenant:** Those households renting part of a dwelling from the occupant household.
- C. **Other:** Those households comprising persons living permanently in rooming houses, persons without dwellings, and persons living abroad although enumerated in the homeland (e.g., diplomats).

FINLAND

3.	Population by age and sex, by single years, by provinces, statistical regions, in Helsinki, Tampere and Turku and Coast Region of Vaasa.	92
4.	Population by age, sex and marital status, by 5 years, by provinces, statistical regions, in Helsinki, Tampere and Turku and Coast Region of Vaasa.	102
5.	Population by main language, by provinces and in Helsinki, Tampere and Turku.	118
6.	Population by congregation, age and sex, in whole country, urban communes and Helsinki, Tampere and Turku.	120
7.	Population by congregation, by provinces.	122
8.	Non-administrative urban settlements with 150-199 inhabitants.	124

III. ECONOMICALLY ACTIVE POPULATION BY INDUSTRY AND INDUSTRY STATUS.

TABLES

Number	Title	Page
1.	Economically active population by industry (3-digit level) and industrial status; whole country, urban communes.	28
2.	Economically active population by industry [2-digit level] and by detailed classification of industrial status; whole country, urban communes.	54
3.	Economically active population by industry (3-digit level), by statistical region.	58

IV. POPULATION BY INDUSTRY AND INDUSTRIAL STATUS.

TABLES

Number	Title	Page
1.	Population by industry [2-digit level] and industrial status, by statistical regions, communes, and non-administrative urban settlements.	2
2.	Population by industry [3-digit level] and industrial status (cross-table), by statistical regions, in Helsinki, Tampere, Turku, and Coast Region of Vaasa.	102

V. FAMILIES.[1]

TABLES

Number	Title	Page
1.	Families by type, by provinces and in Helsinki, Tampere and Turku.	20
2.	Husband-wife families by socio-economic status of head and number of children under 18 and under 7 years of age; whole country, urban and rural communes and non-administrative urban settlement areas in rural communes.	22
3.	Families of Swedish-speaking heads by socio-economic status of head and number of children under 18 and 7 years of age; husband-wife families; whole country and urban communes.	50
4.	Families by socio-economic status of head, housewife's economic activity and number of children under 18 years of age; whole country, urban and rural communes and non-administrative urban settlement areas in rural communes.	66
5.	Husband-wife families in which wife is economically active by socio-economic status of husband and wife; whole country, urban and rural communes and non-administrative urban settlement areas in rural communes.	78

VI. POPULATION BY SOCIO-ECONOMIC STATUS; ECONOMICALLY ACTIVE POPULATION BY AGE, LOCATION OF WORKING PLACE.

TABLES

Number	Title	Page
1.	Population by socio-economic classification.	3
2.	Economically active population by sex, age, and industrial status; statistical regions, Helsinki, Tampere, Turku, and Coast Region of Vaasa.	4
3.	Economically active population by commune of residence and working place, by industry.	28

---

[1]Classification by type of family is as follows: (1) married couples without children, (2) married couples with children, (3) man with children, and (4) woman with children. Persons in common-law marriages are classified as married if they have common children.

VII. HOUSEHOLDS AND THEIR HOUSING CONDITIONS.

TABLES

Number	Title	Page
1.	Households by size, by provinces, statistical regions, communes and non-administrative urban settlements.	2
2.	Households by marital status, age and sex of head and by size of household, by statistical regions, in Helsinki, Tampere and Turku and Coast Region of Vaasa.	29
3.	Occupant households by socio-economic status of head, number of persons in household and number of rooms in their use; whole country, urban communes, and rural communes.	105
4.	Sub-tenant households by socio-economic status of head, number of persons in household and number of rooms in their use; whole country, urban communes, and rural communes.	112
5.	Occupant households by size and number of rooms in their use, by provinces and statistical regions, in Helsinki, Tampere and Turku and Coast Region of Vaasa.	119
6.	Sub-tenant households by size and number of rooms in their use, by provinces and statistical regions, in Helsinki, Tampere and Turku and Coast Region of Vaasa.	134

VIII. POPULATION BY BIRTH-PLACE AND BY EDUCATION, DISPLACED POPULATION, ETC.[1]

TABLES

Number	Title	Page
1.	Population by birth-place, by provinces and statistical regions, in Helsinki, Tampere, Turku and Coast Region of Vaasa.	2

---

[1] The birth-place of a child is recorded as the mother's commune of residence at the time of birth, rather than the place where the birth occurred. With respect to education, the population is divided into (a) persons with neither middle school nor matriculation examination, (b) persons with middle school education, and (c) persons with matriculation examination. The term displaced population refers to those persons whose permanent place of residence on September 1, 1939, had been in the territory subsequently ceded or leased to the U.S.S.R.

TABLE TITLES BY VOLUME

2. Persons born abroad by country of birth, by provinces, in Helsinki, Tampere and Turku. — 11
3. Population by education and age, by provinces, and statistical regions, in Helsinki, Tampere, Turku and Coast Region of Vaasa. — 12
4. Population by congregation and socio-economic status; whole country. — 36
5. Swedish-speaking population by sex and age, by provinces and in Swedish and bilingual communes. — 38
6. Swedish-speaking population by industry and industrial status; whole country, urban and rural communes, provinces of Uusimaa, Turku-Pori, Ahvenanmaa (Aland Islands) and Vaasa and cities of Helsinki and Turku. — 42
7. Aliens by citizenship, age and sex; whole country, urban communes, Helsinki, Tampere and Turku. — 48
8. Displaced population by ceded and leased territories, by provinces. — 50
9. Displaced population by present commune of residence. — 54
10. Displaced population by age and socio-economic status; whole country. — 62

IX. OCCUPATION AND VOCATIONAL TRAINING.

TABLES

Number	Title	Page
1.	Population by vocational training and age; whole country.	28
2.	Population by occupation and age; whole country.	84
3.	Economically active persons with college or university degree classified by type of degree and occupation; whole country.	112
4.	Economically active population by occupation and vocational training (excluding persons with college or university degree); whole country.	120
5.	Economically active population by vocational training and industry; whole country.	132

FINLAND

XI. NON-ADMINISTRATIVE URBAN SETTLEMENTS AND THEIR BOUNDARIES, ETC.

TABLES

Number	Title	Page
1.	Population, area and population density of settlements, number of map on which settlement is situated, and coordinate numbers of boundaries of settlements according to the basic map.	22
2.	Population and number of households and dwelling units by provinces, statistical regions, communes and parts of settlements.	81

XII. SUPPLEMENTARY VOLUME.

TABLES

Number	Title	Page
1.	Population by industry (1-digit level) and industrial status and age; whole country.	2
2.	Population by industry (3-digit level) and industrial status; whole country.	14
3.	Economically active population by industry (1-digit level) and occupation (partly 3-digit level);[1] whole country.	22
4.	Salaried people by industry (2-digit level), industrial status and employer; whole country.	30
5.	Males by age and number of months in logging and floating; whole country, urban and rural communes.	38
6.	Males by industry (2-digit level) and number of months in logging and floating; whole country.	48
7.	Families by type, number of persons and number of children under 18 years of age; by provinces, statistical regions and communes.	51
8.	Man-and-children families and woman-and-children families by socio-economic status and age of head and number of children under 18 years of age; whole country.	78
9.	Woman-and-children families by civil status and age of head and number of children under 18 years of age; whole country.	88

[1] Abbreviated list of occupations at the 3-digit level of classification.

TABLE TITLES BY VOLUME

10. Husband-wife families without children by socio-economic status and age of head; whole country. — 90
11. Households by industry and industrial status of head; whole country, urban and rural communes. — 91
12. Occupant households by age, sex, and civil status[2] of head; whole country, urban and rural communes. — 94
13. Population by status in household, by statistical regions. — 95
14. Dwelling units by number of rooms and area, by provinces. — 98
15. Total areas of dwelling units and area per dwelling unit and persons, by provinces, statistical regions, and communes. — 111

---

[2] Marital status.

F R A N C E

MAIN ENTRIES AND VOLUME (FASCICLE) NOTES:

[Résultats du Dépouillement Exhaustif]

France. Institut National de la Statistique et des Études Économiques.
Recensement général de la population de 1962. Résultats du dépouillement exhaustif. Population par région agricole; fascicules départementaux [Paris, n.d.]
   v. 27cm.

Ain. 45 p.
Aisne. nyp[1]
Allier. 31 p.
Alpes (Basses). nyp
Alpes (Hautes). nyp
Alpes-Maritimes. nyp
Ardèche. 41 p.
Ardennes. 26 p.
Ariège. np[1]
Aube. 42 p.
Aude. 38 p.
Aveyron. np
Bouches-du-Rhône. nyp
Calvados. 29 p.
Cantal. 51 p.
Charente. 56 p.
Charente-Maritime. 44 p.
Cher. 39 p.
Corrèze. 47 p.
Corse. nyp
Côte-d'Or. 37 p.
Côtes-du-Nord. 32 p.
Creuse. 31 p.
Dordogne. 36 p.
Doubs. np
Drôme. 53 p.
Eure. 53 p.
Eure-et-Loire. 31 p.
Finistère. 48 p.

Gard. 46 p.
Garonne (Haute). np
Gers. np
Gironde. 58 p.
Hérault. 42 p.
Ille-et-Vilaine. 40 p.
Indre. 31 p.
Indre-et-Loire. 49 p.
Isère. 31 p.
Jura. 51 p.
Landes. 36 p.
Loir-et-Cher. 51 p.
Loire. 53 p.
Loire (Haute). 43 p.
Loire-Atlantique. 45 p.
Loiret. 47 p.
Lot. np
Lot-et-Garonne. 40 p.
Lozère. 30 p.
Maine-et-Loire. 26 p.
Manche. 37 p.
Marne. 39 p.
Marne (Haute). 42 p.
Mayenne. 23 p.
Meurthe-et-Moselle. 35 p.
Meuse. 27 p.
Morbihan. 28 p.
Moselle. 31 p.
Nièvre. 28 p.

---

[1] Abbreviations: nyp = not yet published; np = lacking page numbers.

Nord. nyp	Seine-Maritime. 33 p.
Oise. nyp	Seine-et-Marne. Région Parisienne I. 61 p.
Orne. 33 p.	
Pas-de-Calais. nyp	Seine-et-Oise; Seine. Région Parisienne II. 126 p.
Puy-de-Dôme. 63 p.	
Pyrénées (Basses). 40 p.	Sèvres (Deux). 56 p.
Pyrénées (Hautes). np	Somme. nyp
Pyrénées-Orientales. 46 p.	Tarn. np
Rhin (Bas). 31 p.	Tarn-et-Garonne. np
Rhin (Haut). 43 p.	Var. nyp
Rhône. 37 p.	Vaucluse. np
Saône (Haute). 43 p.	Vendée. 35 p.
Saône-et-Loire. 43 p.	Vienne. 48 p.
Sarthe. 45 p.	Vienne (Haute). 23 p.
Savoie. 57 p.	Vosges. 39 p.
Savoie (Haute). 77 p.	Yonne. 37 p.
Seine. [See Seine-et-Oise]	Belfort (Territoire de). 27 p.

France. Institut National de la Statistique et des Études Économiques.
Recensement général de la population de 1962. Résultats du dépouillement exhaustif. Population, ménages, logements, immeubles; fascicules départementaux.[1] Paris, 1966.
96 v. 31cm. (67-123463)

1. Ain. 54 p.
2. Aisne. 82 p.
3. Allier. 64 p.
4. Basses-Alpes. 47 p.
5. Hautes-Alpes. 47 p.
6. Alpes-Maritimes. 75 p.
7. Ardèche. 49 p.
8. Ardennes. 63 p.
9. Ariège. 44 p.
10. Aube. 58 p.
11. Aude. 58 p.
12. Aveyron. 57 p.
13. Bouches-du-Rhône. 97 p.
14. Calvados. 73 p.
15. Cantal. 48 p.
16. Charente. 62 p.
17. Charente-Maritime. 71 p.
18. Cher. 60 p.
19. Corrèze. 56 p.
20. Corse. [Never published.]
21. Côte-d'Or. 61 p.
22. Côtes-du-Nord. 58 p.

---

[1] At the time of the 1962 census there were 90 departments in France. On July 10, 1964, however, the departments of Seine and Seine-et-Oise were reorganized into seven departments (marked *) as follows: Essonne, Hauts-de-Seine, Paris, Seine-Saint-Denis, Val de Marne, Val d'Oise, and Yvelines. (See Statesman's Yearbook 1967-68 [New York: St Martin's Press, 1967], pp. 993-995.)
Unlike the tabulations for the agricultural regions, this set of tabulations includes fascicles for both the old and the new departments, as follows: one fascicle each for 89 of the original 90 departments (Corsica was never published), plus one fascicle each for the seven new departments, totaling 96 fascicles. It should be noted that there are two fascicles each numbered "75" and "78."

23. Creuse. 44 p.
24. Dordogne. 64 p.
25. Doubs. 68 p.
26. Drôme. 65 p.
27. Eure. 57 p.
28. Eure-et-Loire. 61 p.
29. Finistère. 72 p.
30. Gard. 72 p.
31. Haute-Garonne. 60 p.
32. Gers. 54 p.
33. Gironde. 75 p.
34. Hérault. 69 p.
35. Ille-et-Vilaine. 65 p.
36. Indre. 55 p.
37. Indre-et-Loire. 56 p.
38. Isère. 71 p.
39. Jura. 60 p.
40. Landes. 57 p.
41. Loir-et-Cher. 52 p.
42. Loire. 80 p.
43. Haute-Loire. 52 p.
44. Loire-Atlantique. 71 p.
45. Loiret. 61 p.
46. Lot. 49 p.
47. Lot-et-Garonne. 53 p.
48. Lozère. 43 p.
49. Maine-et-Loire. 65 p.
50. Manche. 64 p.
51. Marne. 68 p.
52. Haute-Marne. 59 p.
53. Mayenne. 48 p.
54. Meurthe-et-Moselle. 91 p.
55. Meuse. 59 p.
56. Morbihan. 60 p.
57. Moselle. 105 p.
58. Nièvre. 56 p.
59. Nord. 163 p.
60. Oise. 72 p.
61. Orne. 55 p.
62. Pas-de-Calais. 143 p.
63. Puy-de-Dôme. 66 p.
64. Basses-Pyrénées. 68 p.
65. Hautes-Pyrénées. 58 p.
66. Pyrénées-Orientales. 55 p.
67. Bas-Rhin. 70 p.
68. Haut-Rhin. 68 p.
69. Rhône. 71 p.
70. Haute-Saône. 51 p.
71. Saône-et-Loire. 83 p.
72. Sarthe. 57 p.
73. Savoie. 57 p.
74. Haute-Savoie. 60 p.
75. Seine. 84 p.
*75. Paris. 54 p.
76. Seine-Maritime. 85 p.
77. Seine-et-Marne. 70 p.
78. Seine-et-Oise. 94 p.
*78. Yvelines. 56 p.
79. Deux-Sèvres. 53 p.
80. Somme. 70 p.
81. Tarn. 65 p.
82. Tarn-et-Garonne. 47 p.
83. Var. 62 p.
84. Vaucluse. 58 p.
85. Vendée. 52 p.
86. Vienne. 60 p.
87. Haute-Vienne. 55 p.
88. Vosges. 62 p.
89. Yonne. 59 p.
90. Territoire de Belfort. 49 p.
*91. Essonne. 48 p.
*92. Hauts-de-Seine. 42 p.
*93. Seine-Saint-Denis. 46 p.
*94. Val-de-Marne. 46 p.
*95. Val-d'Oise. 55 p.

France. Institut National de la Statistique et des Études Économiques.
Recensement général de la population de 1962. Résultats du sondage au 1/20e pour la France entière: population active. Paris, 1964.
183 p. 27cm. (NUC66-20985)

France. Institut National de la Statistique et des Études
    Économiques.
  Recensement général de la population de 1962. Résultats du
sondage au 1/20e pour la France entière: ménages, familles.
Paris, 1968.
    239 p. 27cm.

France. Institut National de la Statistique et des Études
    Économiques.
  Recensement général de la population de 1962. Résultats du
sondage au 1/20e pour la France entière: logements, immeubles.
Paris, 1965.
    194 p. 27cm. [Excluded.]

France. Institut National de la Statistique et des Études
    Économiques.
  Recensement général de la population de 1962. Résultats du
sondage au 1/20e pour la France entière: structure de la population totale. Paris, 1965.
    129 p. 27cm.   (67-50825)

France. Institut National de la Statistique et des Études
    Économiques.
  Recensement général de la population de 1962. Résultats du
sondage au 1/20e: population, ménages, logements, immeubles;
fascicules régionaux; récapitulation pour la France entière.
Paris, 1964.
    22 v. 31cm.

   F.E. Récapitulation pour la France entière. 1964. 32 p.
   1.1. Région Parisienne (Seine, Seine-et-Marne, Seine-et-
         Oise). 1964. 76 p.
   2.1. Champagne (Ardennes, Aube, Marne, Haute-Marne). 1964.
         68 p.
   2.2. Picardie (Aisne, Oise, Somme). 1964. 60 p.
   2.3. Haute-Normandie (Eure, Seine-Maritime). 1964. 56 p.
   2.4. Centre (Cher, Eure-et-Loir, Indre, Indre-et-Loire,
         Loire-et-Cher, Loiret). 1964. 88 p.
   3.1. Nord (Nord, Pas-de-Calais). 1964. 75 p.
   4.1. Lorraine (Meurthe-et-Moselle, Meuse, Moselle, Vosges).
         1964. 80 p.
   4.2. Alsace (Bas-Rhin, Haut-Rhin). 1964. 56 p.
   4.3. Franche-Comté (Doubs, Jura, Haute-Saône, Territoire
         de Belfort). 1964. 68 p.
   5.1. Basse-Normandie (Calvados, Manche, Orne). 1964. 60 p.
   5.2. Pays de la Loire (Loire-Atlantique, Maine-et-Loire,
         Mayenne, Sarthe, Vendée). 1964. 83 p.
   5.3. Bretagne (Côtes-du-Nord, Finistère, Ille-et-Vilaine,
         Morbihan). 1964. 72 p.

6.1. Limousin (Corrèze, Creuse, Haute-Vienne). 1964. 59 p.
6.2. Auvergne (Allier, Cantal, Haute-Loire, Puy-de-Dôme). 1964. 67 p.
7.1. Poitou-Charentes (Charente, Charente-Maritime, Deux-Sèvres, Vienne). 1964. 63 p.
7.2. Aquitaine (Dordogne, Gironde, Landes, Lot-et-Garonne, Basses-Pyrénées). 1964. 75 p.
7.3. Midi-Pyrénées (Ariège, Aveyron, Haute-Garonne, Gers, Lot, Hautes-Pyrénées, Tarn, Tarn-et-Garonne). 1964. 99 p.
8.1. Bourgogne (Côte-d'Or, Nièvre, Saône-et-Loire, Yonne). 1964. 67 p.
8.2. Rhône-Alpes (Ain, Ardèche, Drôme, Isère, Loire, Rhône, Savoie, Haute-Savoie). 1964. 108 p.
9.1. Languedoc (Aude, Gard, Hérault, Lozère, Pyrénées-Orientales). 1964. 80 p.
9.2. Provence-Côte d'Azur (Basses-Alpes, Hautes-Alpes, Alpes-Maritimes, Bouches-du-Rhône, Var, Vaucluse, Corse). 1964. 99 p.

[Feuilles Récapitulatives Communales Établies par les Mairies]

France. Institut National de la Statistique et des Études Économiques.
Recensement de 1962: population de la France; départements, arrondissements, cantons, et communes. [Paris, 1962]
1141 p. 28cm. (NUC64-70525)

France. Institut National de la Statistique et des Études Économiques.
Recensement de 1962: villes de plus de 5,000 habitants. Paris, 1962.
24 p. 27cm. (NUC65-62799)

France. Institut National de la Statistique et des Études Économiques.
Recensement de 1962: villes et agglomérations urbaines. Paris, 1964.
215 p. 27cm.

France. Institut National de la Statistique et des Études Économiques.
Recensement de 1962: population légale; résultats statistiques. Paris, 1964.
149 p. 27cm. (NUC66-18547)

MAIN ENTRIES AND VOLUME NOTES

France. Institut National de la Statistique et des Études
    Économiques.
    Recensement de 1962: population légale et statistiques communales complémentaires. Paris, 1963.
    3 v.   27cm.   (65-82631)

France. Institut National de la Statistique et des Études
    Économiques.
    Recensement de 1962: les zones de peuplement industriel ou urbain. Paris, 1964.
    133 p.   27cm.   (NUC63-17025)

France. Institut National de la Statistique et des Études
    Économiques.
    Recensement de la population de mars 1962. Population, logements, immeubles des communes rurales classées par canton.
    2 v.   27cm.   [Excluded; computer print-out pages.]

France. Institut National de la Statistique et des Études
    Économiques.
    Recensement de la population de mars 1962. Population, logements, immeubles des communes rurales classées par catégorie de commune 1954.
    2 v.   27cm.   [Excluded; computer print-out pages.]

ENGLISH TRANSLATION OF CENSUS TITLES:

[The Complete Enumeration]

General census of the population of 1962.
    Results of the complete enumeration. Population by
      agricultural region; departmental fascicles. [90
      fascicles]
    Results of the complete enumeration. Population, households,
      individual dwellings, apartment houses; departmental
      fascicles. [96 fascicles]
    Results of a 5 percent sample tabulation for all of France:
      economically active population.
    Results of a 5 percent sample tabulation for all of France:
      households, families.
    Results of a 5 percent sample tabulation for all of France:
      individual dwellings and apartment houses.
    Results of a 5 percent sample tabulation for all of France:
      structure of the total population.
    Results of a 5 percent sample tabulation: population, house-
      holds, individual dwellings, apartment buildings; regional
      fascicles; summary for all of France. [22 fascicles]

FRANCE

[The Population Count Derived from Communal Town Hall Schedules]

Census of 1962:

    Population of France; departments, districts, cantons, and communes.
    Cities of more than 5,000 inhabitants.
    Cities and urban agglomerations.
    De jure population; statistical results.
    De jure population and supplementary statistics for the communes.
    Industrial and urban zones of population.
    Population, dwellings, and apartment houses of rural communes classified by canton. [Two volumes of computer print-out.]
    Population, dwellings, and apartment houses of rural communes classified by category of commune (1954 definition). [Two volumes of computer print-out.]

ENUMERATION DATE: March 7, 1962.

ENUMERATION BASIS: De jure.

GLOSSARY

ADMINISTRATIVE, GEOGRAPHICAL, AND POLITICAL UNITS

Level	English Translation	Vernacular Term
National	French Republic	République française
Intermediate	Department	Département
	District	Arrondissement
	Canton	Canton
Local	Commune	Commune
	Administrative seat of commune	Chef-lieu
Population cluster ("locality") concepts:[1]		
	Urban commune	Commune urbain
	Multi-communal agglomeration	Agglomération urbain multi-communale

---

[1] For a discussion of this concept, see European Population Censuses: The 1960 Series, U.N. Doc.: ST/CES/3 (1964), p. 108.

GLOSSARY

Separate town	Ville isolée
Rural commune	Commune rurale

Other concepts:

Categories of communes	Catégories des communes
Zone of industrial or urban population	Zone de peuplement industriel ou urbain

The French Republic (i.e., metropolitan France), at the time of the 1962 census, was composed of 90 <u>departments</u>. On July 10, 1964, however, the number of departments was increased to 95, reflecting a division of the Region of Paris (see footnote 1, p. 55).

The departments consist of <u>districts,</u> and the districts of <u>cantons.</u>

<u>Communes</u> are the units of local government, and they vary greatly in size and population. In 1962 there was an average of 12 communes in a canton. However, some of the larger communes, individually, made up several cantons.[2]

For statistical purposes, communes are classed according to population clustering or agglomerations. <u>Urban communes</u> have at least 2,000 inhabitants who are settled in houses no more than 200 meters apart. Such communes may be urban agglomerations or towns. If a population agglomeration of more than 2,000 people extends over several communes, the communes in their totality constitute a <u>multi-communal agglomeration.</u> If a population agglomeration of more than 2,000 people is situated in a single commune, the commune is considered urban in its totality and constitutes a <u>separate town.</u> Towns and agglomerations are categorized according to the size of their populations. All communes which are not urban communes are <u>rural communes.</u>

<u>Categories of communes</u> vary widely in the numerous tables of the census. Communes are combined for tabulation into from two to nine different categories, as follows:

Two categories: Rural communes; and cities and urban agglomerations.

Three categories: Rural communes; cities and agglomerations of less than 100,000; and cities and agglomerations of more

---

[2]<u>Statesman's Yearbook 1967-68</u> (New York: St. Martin's Press, 1967), p. 994.

than 100,000 or the residential complex of the Parisian agglomeration.

Four categories: Rural communes; cities and agglomerations of less than 100,000; cities and agglomerations of more than 100,000; and the residential complex of the Parisian agglomeration.

Five categories: Rural communes; expanding rural communes; cities and agglomerations of less than 100,000; cities and agglomerations of more than 100,000; and the residential complex of the Parisian agglomeration.

Eight categories: Rural communes; towns and agglomerations of less than 5,000; towns and agglomerations of from 5,000 to 10,000; towns and agglomerations of from 10,000 to 20,000; towns and agglomerations of from 20,000 to 50,000; cities and agglomerations of from 50,000 to 100,000; cities and agglomerations of more than 100,000; and the residential complex of the Parisian agglomeration.

Nine categories: Rural communes; total of cities and urban agglomerations; towns and agglomerations of less than 5,000; towns and agglomerations of from 5,000 to 10,000; towns and agglomerations of from 10,000 to 20,000; towns and agglomerations of from 20,000 to 50,000; cities and agglomerations of from 50,000 to 100,000; cities and agglomerations of more than 100,000; and the residential complex of the Parisian agglomeration.

Zones of industrial or urban population are areas intermediate in size between urban agglomerations and purely rural zones. They are marked by three characteristics:

A. Homogeneity of population; specifically, a small proportion of farmers.

B. Mobile labor force.

C. Economic activity created because of proximity to a large city, natural resources, and the existence of adequate transportation for the labor force.

TABLE OF DEMOGRAPHIC CONCEPTS

The demographic concepts which appear in the following table correspond to the recommendations of the European Conference of Statisticians, unless otherwise noted. They are defined and discussed in the appendixes and in footnotes. Terms in parentheses are alternative forms which appear in the table titles. Terms in brackets are not used in the table titles, but they are included here in order to give logical completeness to the various classifications in the table.

GLOSSARY

Appendix	English Table Term	Vernacular Table Term
	CULTURAL, EDUCATIONAL, AND PERSONAL CHARACTERISTICS	
...	Age[1]	Âge
	ECONOMIC CHARACTERISTICS	
A	Economic activity	Activité économique
	Economically active	Active
	Employed	Ayant un emploi
	Unemployed	Sans emploi
	Economically inactive	Inactif
B	Employment status	Statut
C	Industrial classification	Catégorie d'activité économique
	Division(s)	Branches (regroupements par branches)
	Major group(s)	Groupes (regroupements en 41 postes or 2 chiffres de la nomenclature officielle)
	[Group]	...
	Subgroup(s)	Activité économique détaillée
D	[Occupational classification][2]	...
	Major group(s)	Activité individuelle
E	Socio-economic category[3]	Catégorie socio-économique
	HOUSEHOLD AND FAMILY CHARACTERISTICS	
F	Household	Ménage
	Private	Ordinaire
	Institutional	Collectif
	Family	Famille
	Nuclear	Noyau
	Biological[4]	Biologique
	Marital status[5]	État matrimonial
	POPULATION DEFINITIONS: TOWN HALL COUNTS[6]	
...	Category of population[7]	Catégorie de population
	De jure population	Population légale
	Total municipal population	Population municipale totale

---

[1] [All footnotes are at the end of the Table.]

FRANCE

Population enumerated separately	Population comptée à part
Total population	Population totale
Population enumerated separately which belongs to the municipal population of other communes (called the "double count")	Population comptée à part appartenant à la population municipale d'autres communes (doubles comptes)
Total population without the double count in the population enumerated separately	Population totale sans doubles comptes dans la population comptée à part

---

[1] Age is that attained in 1962 (derived from data on year of birth).

[2] The classification by occupation can be converted to the ISCO 1958 at the major group level (1-digit) only. In the French table titles, however, the term "activité individuelle" is further broken down into two categories, "rubriques détaillées" and "groupes à 2 chiffres." While these categories provide a detailed classification of occupations, they are not consistent with the ISCO 1958 recommended schema.

[3] The classification by socio-economic category consists of two digit levels: the 1-digit level referring to <u>groups</u>, and the 2-digit level referring to <u>categories</u>. The 10 groups, numbered 0-9, are divided into a total of 37 categories. The left-hand digit of the category specifies the group to which it belongs.

[4] The biological family is the aggregate of persons of the same nuclear family formed by the head of the family, his or her spouse, and his or her children not married and younger than 25 years.

[5] In the table titles, "marital status" may distinguish between married and single persons, or it may distinguish among married, single, widowed, and divorced. Usage varies.

[6] The Synoptic Table of Population Definitions on the following page presents definitions for all the population terms employed in that part of the census derived from the Town Hall Counts:

GLOSSARY

SYNOPTIC TABLE OF POPULATION DEFINITIONS

Term	1962 Definition	1954 Definition
	(See "Components of Definitions" below)	
Total municipal population	A + B + C + H	A
Population enumerated separately	D + E + G	C + D + E + G
Total population	(A + B + C + H) + (D + E + G)	A + (C + D + E + G)
De jure population	(A + B + C + H) − D	
Population enumerated separately which belongs to the municipal population of other communes (double count)	D	
Total population without the double count in the population enumerated separately	(A + B + C + H) + (E + G)	

COMPONENTS OF DEFINITIONS USED IN SYNOPTIC TABLE

Persons whose private residence is within the commune | Persons residing in institutions of the commune

A. Persons who live at their own private residence

Military personnel, students, etc. who: | Military personnel, students, etc. who:

B. Reside in an institution located in another commune

C. Reside in an institution of the commune

C. Have a private residence in the commune

D. Have a private residence in some other commune

E. Do not have a private residence

F. Persons confined in an institution (e.g., psychiatric hospital) in the commune or in some other commune

G. Persons confined in an institution (e.g., psychiatric hospital) in the commune who have or do not have a private residence in some other commune

H. Military personnel on foreign duty

---

[7] The term "category of population" refers to the population in private households, as opposed to the population in institutional households (by the 1962 and 1954 definitions).

FRANCE

TABLE TITLES BY VOLUME

GENERAL CENSUS OF THE POPULATION OF 1962. RESULTS OF THE COMPLETE ENUMERATION. POPULATION BY AGRICULTURAL REGION; DEPARTMENTAL FASCICLES.
[90 fascicles; see listing on pp. 54-55 above.]

Structure of the Fascicles

The 90 fascicles are organized by departments and by agricultural regions within departments.[1] Basically, each fascicle consists of two sections: (1) a set of tables for the entire department, numbered 1-14, followed by (2) a set of tables, numbered 1-9, for each agricultural region within the department. The first 9 table titles of the departmental sections are identical to the first 9 table titles for the agricultural sections. Table titles 10-14 of the departmental sections are not repeated in the agricultural section.

The fascicles vary somewhat, however, in the materials included. Tables of contents are not uniformly included, and page locations for the same tables are not the same from fascicle to fascicle. Entire tables and/or pages may be omitted from a particular fascicle. As far as we can ascertain, some fascicles have not been published to date. Because of lack of uniformity of the fascicles, page locations have been omitted.

TABLES

Number	Title
1.	Population in private households by sex, age, and whether or not the head of the household is in agriculture.
2.	Total population by sex and category (whether in or outside a private household).
3.	Economically active population by sex, age, and whether or not the head of the household is in agriculture.
4.	Total population by sex and socio-economic category.

---

[1] For a complete list of agricultural regions arranged alphabetically and also by department, see: France, I.N.S.E.E., Recensement général de l'agriculture de 1955 (Paris: Imprimerie Nationale, 1959).

5. Economically active agricultural population by age, sex, and employment status.
6. Employed economically active persons by sex and industrial classification.
7. Households by socio-economic category of the head of the household.
8. Farming by specialization and area.
9. Farms and agricultural buildings by the year of construction and water supply.
10. Changes in the agricultural population by employment status from 1954 to 1962.
11. Changes in selected characteristics of the economically active agricultural population by age from 1954 to 1962.
12. Economically active agricultural population who are foreigners.
13. Net outmigration of the economically active male agricultural population by generation from 1954 to 1962.
14. Changes in the economically active agricultural population by agricultural region from 1954 to 1962.

GENERAL CENSUS OF THE POPULATION OF 1962. RESULTS OF THE COMPLETE ENUMERATION. POPULATION, HOUSEHOLDS, INDIVIDUAL DWELLINGS, APARTMENT HOUSES; DEPARTMENTAL FASCICLES. [96 fascicles; see listing on pp. 55-56 above.]

Structure of the Fascicles

There are five categories of table titles in the departmental fascicles which are labeled as follows:

Number	Category	Table Titles Numbers
I.	Departmental Tables	D 1 through D 33
II.	Large Agglomerations	GA 1 through GA 19
III.	Agglomerations	A 1 through A 12
IV.	Rural Communes and Urban Agglomerations of Each Canton	V 1 through V 7
V.	All Communes	C (One table only)

FRANCE

Each departmental fascicle contains tables for categories I, IV, and V. However, some departments contain no agglomerations and/or no large agglomerations, and consequently do not have tables for Categories II and/or III. The following list indicates which fascicles do not have tables for Category II and/or Category III:

Number	Department	Omits II	Omits III	Number	Department	Omits II	Omits III
1.	Ain	X		40.	Landes	X	
4.	Basses-Alpes	X	X	41.	Loir-et-Cher	X	
5.	Hautes-Alpes	X		42.	Loire		X
7.	Ardèche	X	X	43.	Haute-Loire	X	
9.	Ariège	X	X	46.	Lot	X	X
10.	Aube		X	47.	Lot-et-Garonne	X	
11.	Aude	X		48.	Lozère	X	X
12.	Aveyron	X		49.	Maine-et-Loire	X	X
15.	Cantal	X		50.	Manche		X
19.	Corrèze	X		52.	Haute-Marne	X	
21.	Côte-d'Or		X	53.	Mayenne	X	
22.	Côtes-du-Nord		X	55.	Meuse	X	
23.	Creuse	X	X	58.	Nièvre		X
25.	Doubs		X	61.	Orne	X	
27.	Eure	X		63.	Puy-de-Dôme		X
31.	Haute-Garonne		X	64.	Basses-Pyrénées		X
32.	Gers	X		65.	Hautes-Pyrénées		X
36.	Indre		X	66.	Pyrénées-Orientales		X
37.	Indre-et-Loire		X	68.	Haut-Rhin		X
39.	Jura	X					

TABLE TITLES BY VOLUME

Number	Department	Omits II	Omits III	Number	Department	Omits II	Omits III
70.	Haute-Saône	X	X	85.	Vendée	X	
72.	Sarthe		X	87.	Haute-Vienne		X
73.	Savoie		X	88.	Vosges	X	
75.	Seine		X	89.	Yonne	X	
75A.	Paris	X	X	90.	Territoire de Belfort		X
77.	Seine-et-Marne	X		91.	Essonne	X	X
78.	Seine-et-Oise	X		92.	Hauts-de-Seine	X	X
78A.	Yvellines	X	X	93.	Seine-Saint-Denis	X	X
79.	Deux-Sèvres	X					
81.	Tarn	X		94.	Val-de-Marne	X	X
82.	Tarn-et-Garonne	X		95.	Val-d'Oise	X	X

TABLES[1]

Number	Title

I. Departmental Tables

D 1. Total population by sex, age, and marital status.

D 2. Total population by category (1962 and 1954 definitions), sex, and age.

D 3. Economically active population by sex, age, agricultural and nonagricultural activity, and employment status.

D 4. Total population and economically active population of agricultural and nonagricultural households by sex, age, and employment status.

D 5. Total economically active population, agricultural and nonagricultural, by sex and age.

D 6. Employed economically active population by sex, industrial major group, and employment status.

D 7. Economically active population by sex and detailed employment status.

D 8. Employed economically active population by sex, industrial major group (1959 definition),[2] and employment status.

---

[1] No page numbers provided in original text.

[2] The 1959 definitions of industrial classification (which are

D 9. Economically active population by sex, socio-economic category, and employment status.

D 10. Economically inactive population 15 years of age and over by sex and socio-economic category.

D 11. Employed foreigners by sex and industrial division.

D 12. Economically active foreigners by sex and socio-economic category.

D 13. Naturalized citizens by country of origin; foreigners by nationality.

D 14. Naturalized citizens and foreigners by sex and age.

D 15. Institutional households, and population enumerated separately by department (1954 definition), by sex.

D 16. Pupils or students and apprentices 15 years of age and over by sex, age, and category of commune.

D 17. Pupils or students 11 years of age and over by subject of study, sex, age, and category of commune.

D 18. Non-student population 15 years of age and over (excluding apprentices) by sex, age, and type of their stated general educational diploma.

D 19. Non-student population (excluding apprentices) by sex according to the type of their stated professional or technical diploma.

D 20. Population in private households by socio-economic category of economically active persons and by social status of the head of the household for economically inactive persons. Population by socio-economic category of the head of the household.

D 21. Analysis of the composition of private households by category of commune.

D 22. Private households by number of persons and by sex, marital status, and age of the head of the household.

D 23. Private households by number of persons and socio-economic category of the head of the household.

D 24. Biological families by the number of children 16 years of age and under and the socio-economic category of the head of the household.

---

the definitions employed in the 1962 census) are not precisely comparable with the definitions of 1954 beyond the Industrial Division (branche d'activité économique), or 1-digit, level. For a full discussion of how these definitions differ, consult the introductory material to the 1962 census volume entitled Résultats du sondage au 1/20e pour la France entière: population active (Paris: Direction des Journaux Officiels, 1964).

D 25. Distribution of principal residences by the number of rooms, by category of commune.

D 26. Distribution of private dwellings by kind of kitchen and number of rooms, by category of commune.

D 27. Characteristics and occupancy status of principal residences[3] by category of commune.

D 28. Distribution of principal residences[3] by period of construction and category of commune.

D 29. Distribution of principal residences[3] by socio-economic category of the head of the household and occupancy status.

D 30. Distribution of principal residences[3] by socio-economic category of the head of the household, occupancy status, and the index of population.[4]

D 31. Combined characteristics of occupied dwellings by category of dwelling and number of rooms (entire department).

D 31A. Combined characteristics of occupied dwellings by category of dwelling and number of rooms (towns and urban agglomerations).

D 32. Apartment buildings by type and category of commune.

D 33. Characteristics of apartment buildings with at least one occupied dwelling by category of commune.

II. Large Agglomerations

GA 1. Total population by sex, age, and marital status.

GA 2. Total population by category (1954 and 1962 definitions), sex, and age.

GA 3. Pupils or students and apprentices 15 years of age and over by sex and age.

GA 4. Pupils or students 11 years of age and over by subject of study, sex, and age.

---

[3] A principal residence is one category of lodging; others are vacant buildings and secondary residences. In this category, the 1962 census has included all types of private habitation which are separate and used as dwellings (e.g., shanties used as private dwellings). The place must be closed off by walls and partitions, must not be connected by halls with another building, and must actually be used for habitation.

[4] The index of population of dwellings is a comparison of the actual density of dwellings with pre-established norms for the number of persons per room. This comparison provides a scale ranging from "accentuated underpopulation" to "accentuated overpopulation," on which each principal residence is rated.

GA 5. Economically active population by sex, age, and employment status.

GA 6. Employed economically active population by sex, industrial major group, and employment status.

GA 7. Economically active population by sex and detailed employment status.

GA 8. Economically active population by sex, socio-economic category, and employment status.

GA 9. Economically inactive population 15 years of age and older by sex and socio-economic category.

GA 10. Private households by number of persons, by the sex, marital status, and age of the head of the household.

GA 11. Private households by number of persons, by the socio-economic category of the head of the household.

GA 12. Analysis of the composition of private households.

GA 13. Distribution of principal residences[3] by number of rooms and number of occupants.

GA 14. Characteristics and occupancy status of principal residences.[3]

GA 15. Distribution of private dwellings by kind of kitchen and number of rooms.

GA 16. Allotment of principal residences[3] by period of construction.

GA 17. Combined characteristics of occupied dwellings by category of dwelling and number of rooms.

GA 18. Distribution of principal residences[3] by the socio-economic category of the head of the household, occupancy status, and index of population.[4]

GA 19. Number of apartment buildings. Characteristics of apartment buildings with at least one occupied dwelling.

III. Agglomerations

A 1. Total population by sex and age.

A 2. Economically active population by sex and detailed employment status.

A 3. Employed economically active population by sex, industrial major group, and employment status.

A 4. Total population by sex, socio-economic category, and employment status. Private households by the socio-economic category of the head of the household.

A 5. Private households by number of persons, by the socio-economic category of the head of the household.

TABLE TITLES BY VOLUME

A 6. Analysis of the composition of private households.

A 7. Distribution of principal residences[3] by number of rooms and number of occupants.

A 8. Characteristics and occupancy status of principal residences.[3]

A 9. Distribution of private dwellings by number of rooms and kind of kitchen.

A 10. Distribution of principal residences[3] by period of construction.

A 11. Combined characteristics of occupied dwellings by category of dwelling.

A 12. Characteristics of apartment dwellings with at least one occupied dwelling.

IV. Rural Communes and Urban Agglomerations of Each Canton

V 1. Total population by sex and age. Foreigners (including Muslims originally from Algeria).

V 2. Population of institutional households. Private households and the population of private households by the socio-economic category of the head of the household.

V 3. Economically active population by sex and socio-economic category.

V 4. Economically active population by sex and employment status. Economically inactive population 15 years of age and older by sex and socio-economic category.

V 5. Private households by number of persons. Biological families. Independent workers in private households. Places of residence by occupancy status. Overpopulation of principal residences.[3]

V 6. Private dwellings by number of rooms and period of construction. Other principal places of residence.[3] Vacant dwellings. Secondary places of residence.[5]

V 7. Private dwellings by basic kinds of facilities.

V. All Communes

C. Total population. Population of private households. Private households. Private dwellings.

---

[5]Secondary residences are furnished dwellings rented, or for rent, at seaside or winter sports resorts, etc.

FRANCE

GENERAL CENSUS OF THE POPULATION OF 1962. RESULTS OF A 5 PERCENT SAMPLE FOR ALL OF FRANCE: ECONOMICALLY ACTIVE POPULATION

TABLES

Number	Title	Page
1.	Economically active population by sex, socio-economic category, and employment status.	66
2.	Employed economically active population by industrial major group, sex, and employment status.	72
3.	Employed economically active population by industrial sub-group, sex, and employment status.	78
4.	Employed economically active population by industrial major group, sex, and employment status.	116
5.	Economically active population by trade or occupational major group, sex, and employment status.	122
6.	Economically active population by trade or occupational major group, sex, and employment status.	142
7.	Employed economically active population by industrial major group, sex, and socio-economic category (22 headings).	144
8.	Economically active population, agricultural or nonagricultural, by sex, age, and employment status.	150
9.	Total population and economically active population (by employment status) in agricultural and nonagricultural households by sex and age.	154
10.	Population 15 to 24 years of age: apprentices, other economically active categories; the contingent military; pupils and students; economically inactive who are not in school.	159
11.	Economically active female population by age and marital status.	160
12.	Economically active agricultural and nonagricultural female population by marital status.	160
13.	Economic activity of married women (living with their husbands) by age and family status (number and age of children).	161
14.	Economic activity of married women (living with their husbands) by age and socio-economic category of the husband.	162
15.	Economically active population by socio-economic category, sex, and age.	164
16.	Employed economically active population by industrial major group, sex, and age.	166

TABLE TITLES BY VOLUME

17.	Foreigners and Algerians by sex, socio-economic category, and employment status.	174
18.	Foreigners (grouped according to principal nationalities) by sex, socio-economic category, and employment status.	176
19.	Foreigners and Algerians by sex, industrial major group, and employment status.	179
20.	Foreigners (grouped according to principal nationalities) by sex and industrial major group.	180
21.	Unemployed by sex, age, and duration of time seeking employment.	182
22.	Unemployed by sex and occupation.	183

GENERAL CENSUS OF THE POPULATION OF 1962. RESULTS OF A 5 PERCENT SAMPLE FOR ALL OF FRANCE: HOUSEHOLDS, FAMILIES.[1]

TABLES

Number	Title	Page
	A. Households	
1.	Private households by number of persons (6 categories), by the sex, age group (7 categories), and marital status of the head of the household:	
	Part 1: Number of persons 1962 definition.	68
	Part 2: Number of persons 1954 definition.	69
2.	Private households by number of persons (6 categories), by the socio-economic category (37 and 10 categories) of the head of the household, by category of commune (4 categories):	
	Part 1: Number of persons 1962 definition.	70
	Part 2: Number of persons 1954 definition.	72

---

[1]There are three kinds of family nuclei (noyaux):

1. Single nucleus (noyau unique) in a household
2. Secondary nucleus (noyau secondaire) ⎫ Types of nuclei in a
3. Principal nucleus (noyau principal) ⎭ multi-family household.

The term single nucleus refers to a one-family household. The term secondary nucleus refers to a nucleus in a household whose head belongs to another nucleus within the household. The term principal nucleus refers to the aggregate of persons (including the household head) who are not in the secondary nucleus (or nuclei). The composition of each of these three types may correspond to (a), (b), or (c) as discussed under family nuclei in Appendix F.

FRANCE

3. Private households by number of persons (6 categories), by the socio-economic category (2 categories) of the head of the household, by category of commune (8 categories):
   Part 1:   Number of persons 1962 definition.   74
   Part 2:   Number of persons 1954 definition.   75

4. Population in private households (1962 definition) by individual socio-economic category and sex, by the socio-economic category of the head of the household.   76

5. Population in private households by the socio-economic category (10 categories) of the household members as indicated by that of the head of the household:
   Part 1:   Population in private households 1962 definition.   84
   Part 2:   Population in private households 1954 definition.   86

6. Private households by total number of persons (6 categories, excluding domestic servants and employees living in the household) and number of economically active persons in the household, by the socio-economic category (37 and 10 categories) of the head of the household.   88

6A. Private households by number of economically active persons in the household (5 categories), by the socio-economic category (37 and 10 categories) of the head of the household.   91

7. Private households by total number of persons (12 categories, excluding domestic servants and employees living in the household) and number of children 16 years of age and younger, by the socio-economic category (18 categories) of the head of the household.   92

7A. Private households by number of children 16 years of age and younger (7 categories), by the socio-economic category of the head of the household.   97

8. Private households by total number of persons (12 categories, excluding domestic servants and employees living in the household) and number of children 16 years of age and younger, by the sex, age group (9 categories), and marital status (2 categories) of the head of the household.   98

8A. Private households by number of children 16 years of age and younger (7 categories), by the sex, age group (9 categories), and marital status (2 categories) of the head of the household.   103

9.	Private households by the sex, single years of age, and marital status (4 categories) of the head of the household.	104
10.	Population in private households (except mobile dwellings) by the sex, age (9 categories), and relationship to the head of the household (9 categories), by category of commune (4 categories).	106
11.	Private households by family nuclei and biological families, by the socio-economic category (2 categories) and age group (6 categories) of the head of the household, by category of commune (4 categories).	109
12.	Private households (family nucleus structure) by the socio-economic category (2 categories) and age group (6 categories) of the head of the household, by category of commune (5 categories).	112
13.	Private households (biological family structure) by the socio-economic category (2 categories) and age group (6 categories) of the head of the household, by category of commune (5 categories).	116

B. Family Nuclei

14.	Nuclei (single, principal, and secondary) by the sex of the head of the nucleus and number of children (5 categories) by type of nucleus:	
	(a) Total nuclei.	122
	(b) Nuclei headed by men.	124
	(c) Nuclei headed by women.	126
15.	Part 1: Types of nuclei by the socio-economic category (2 categories) and age group (6 categories) of the head of the nucleus, by category of commune (4 categories):	
	(a) Single nuclei, each composed of one couple, headed by a man.	128
	(b) Single nuclei, each composed of one couple, headed by a woman.	135
	(c) Single nuclei, not couples, headed by men.	136
	(d) Single nuclei, not couples, headed by women.	139
	Part 2: Types of principal nuclei by the socio-economic category (2 categories) and age group (6 categories) of the head of the nucleus, by category of commune (4 categories):	
	(a) Principal nuclei, each composed of one couple, headed by a man.	143
	(b) Principal nuclei, each composed of one couple, headed by a woman.	145
	(c) Principal nuclei, not couples, headed by men.	146
	(d) Principal nuclei, not couples, headed by women.	147

Part 3: Types of secondary nuclei by the socio-
economic category (2 categories) and age group (6
categories) of the head of the nucleus, by category
of commune (4 categories):
(a) Secondary nuclei, each composed of one couple,
headed by a man. 148
(b) Secondary nuclei, each composed of one couple,
headed by a woman. 150
(c) Secondary nuclei, not couples, headed by men. 151
(d) Secondary nuclei, not couples, headed by women. 152

16. Part 1: Single nuclei by the sex, age group (6 categories), marital status (2 categories), and socio-
economic category (2 categories) of the head and by
category of commune (4 categories), taking into account whether or not the nucleus includes a
biological family:
(a) Total single nuclei. 153
(b) Single nuclei headed by men. 154
(c) Single nuclei headed by women. 155

Part 2: Principal nuclei by the sex, age group (6
categories), marital status (2 categories), and
socio-economic category (2 categories) of the head
and by category of commune (4 categories), taking
into account whether or not the nucleus includes
a biological family:
(a) Total principal nuclei. 156
(b) Principal nuclei headed by men. 157
(c) Principal nuclei headed by women. 158

Part 3: Secondary nuclei by the sex, age group (6
categories), marital status (2 categories), and socio-
economic category (2 categories) of the head and by
category of commune (4 categories), taking into account whether or not the nucleus includes a biological
family:
(a) Total secondary nuclei. 159
(b) Secondary nuclei headed by men. 160
(c) Secondary nuclei headed by women. 161

17. Combined characteristics of secondary nuclei and
principal nuclei. 163

18. Nuclei by the sex, single years of age, and marital
status (4 categories) of the head of the nucleus. 165

C. Biological Families

19. Types of biological families by the age group (6
categories) of the head of the family. 167

20. Part 1: Types of biological families (single, principal,
and secondary) by the number of children and socio-
economic category (2 categories) of the head of the
family, by category of commune (4 categories):

	(a)	Total biological families.	168
	(b)	Biological families headed by men.	170
	(c)	Biological families headed by women.	172

Part 2: Types of single biological families by the number of children and socio-economic category (2 categories) of the head of the family, by category of commune (4 categories):
- (a) Total single biological families. 174
- (b) Single biological families headed by men. 176
- (c) Single biological families headed by women. 178

Part 3: Types of principal biological families by the number of children and socio-economic category (2 categories) of the head of the family, by category of commune (4 categories):
- (a) Total principal biological families. 180
- (b) Principal biological families headed by men. 182
- (c) Principal biological families headed by women. 184

Part 4: Types of secondary biological families by the number of children and socio-economic category (2 categories) of the head of the family, by category of commune (4 categories):
- (a) Total secondary biological families. 186
- (b) Secondary biological families headed by men. 188
- (c) Secondary biological families headed by women. 190

21. Part 1: Types of single biological families by the socio-economic category (2 categories) and age group (6 categories) of the head of the family, by category of commune (4 categories):
- (a) Single biological families, each composed of one couple, headed by a man. 192
- (b) Single biological families, each composed of one couple, headed by a woman. 195
- (c) Single biological families, not couples, headed by men. 196
- (d) Single biological families, not couples, headed by women. 197

Part 2: Types of principal biological families by the socio-economic category (2 categories) and age group (6 categories) of the head of the family, by category of commune (4 categories):
- (a) Principal biological families, each composed of one couple, headed by a man. 199
- (b) Principal biological families, each composed of one couple, headed by a woman. 201
- (c) Principal biological families, not couples, headed by men. 202
- (d) Principal biological families, not couples, headed by women. 203

Part 3: Types of secondary biological families by the socio-economic category (2 categories) and age group (6 categories) of the head of the family, by category of commune (4 categories):
- (a) Secondary biological families, each composed of one couple, headed by a man. ... 204
- (b) Secondary biological families, each composed of one couple, headed by a woman. ... 206
- (c) Secondary biological families, not couples, headed by men. ... 207
- (d) Secondary biological families, not couples, headed by women. ... 208

22. Biological families by number of children 16 years of age and younger, 6 years of age and younger, and 2 years of age and younger; by the sex, age group (9 categories), and marital status (4 categories) of the head of the family:
    - (a) Total biological families. ... 210
    - (b) Biological families headed by men. ... 212
    - (c) Biological families headed by women. ... 213

23. Biological families by number and age of children (3 categories) 16 years of age and younger, by the socio-economic category (2 categories) and employment status (7 categories) of the head of the family. ... 214

24. Biological families by number and age group of children (3 categories)[2] 16 years of age and younger, by the socio-economic category (2 categories) and employment status (3 categories) of the head of the family. ... 216

25. Biological families by number of children two years of age and younger, 6 years of age and younger, and 16 years of age and younger; by the socio-economic category (2 categories) and employment status (2 categories) of the head of the family. ... 219

26. Biological families by the sex, single years of age, and marital status (4 categories) of the head of the family. ... 220

27. Population in biological families by sex, age group (14 categories), relationship to the head of the family (3 categories),[3] by category of commune (5 categories):
    - (a) Total biological families. ... 222
    - (b) Single or principal biological families. ... 224
    - (c) Secondary biological families. ... 226

---

[2] 0-2, 3-6, 7-16 years of age.

[3] Head of the family; spouse of the head of the family; child of the head of the family.

28.	Population in private households outside of biological families by sex, age group (14 categories), and relationship to the head of the household (7 categories), by category of commune (5 categories).	228
29.	Population in private households outside of biological families (persons 15 to 24 years of age living alone) by sex, single years of age, kind of economic activity (5 categories), relationship to the head of the household, by category of commune (4 categories):	
	(a) Heads of household.	232
	(b) Relatives of the head of the household.	234
	(c) Friends of the head of the household.	236
	(d) Boarders or subtenants of the head of the household.	238

GENERAL CENSUS OF THE POPULATION OF 1962. RESULTS OF A 5 PERCENT SAMPLE FOR ALL OF FRANCE: STRUCTURE OF THE TOTAL POPULATION.

TABLES

Number	Title	Page
	Retrospective Tables	
R 1.	Population by sex and age group from 1776 to 1962.	50
R 2.	Population by sex and single years of age from 1954 to 1961.	58
R 3.	Population by sex, age, and marital status.	66
R 4.	Population by nationality from 1851 to 1962.	68
	1962 Census Tables	
1.	Total population: sex, age, and marital status.	78
2.	Total population: sex, age, and marital status by category of commune.	82
3.	Total population: sex, age, and nationality.	87
4.	Total population: sex, marital status, and nationality.	87
5.	French natives: sex and country of birth.	88
6.	Naturalized persons: sex and nationality of origin.	88
7.	Foreigners: sex and nationality.	88
8.	Total population: sex and category of population by category of commune.	90
9.	Total population, population in private households,	

	and population not included in private households: sex and age.	92
10.	Total population, population in private households, and population not included in private households: sex and age by category of commune.	96
11.	Population in private households (1962 definition): sex, age, and marital status.	105
12.	Population in private households (1954 definition): sex, age, and marital status.	105
13.	Population in private households (1962 definition): sex, age, and marital status by category of commune.	106
14.	Population in communities: sex, age, and category of population.	108
15.	Population counted apart and persons in the military stationed outside the capital.	110
	15.1. Military personnel in barracks and camps (men): age, marital status, and category of population.	110
	15.2. Boarders: age and category of population.	110
	15.3. Persons undergoing treatment in sanatoriums, preventoriums, aeriums: sex, age, marital status, and category of population.	111
	15.4. Temporary workers in work camps (men): age, marital status, and category of population.	112
	15.5. Confined persons: sex, age, and marital status.	112
	15.6. Persons undergoing treatment in psychiatric hospitals: sex, age, and marital status.	112
16.	Population in collective households.	113
	16.1. Population in homes for the aged: sex, age, and marital status.	113
	16.2. Population in religious communities: sex and age.	113
	16.3. Population in other collective households: sex, age, and marital status.	113
17.	Total population: broad age groups by department and region.	114
18.	Total population: nationality by department and region.	118
19.	Total population: category of population by department and region.	122

TABLE TITLES BY VOLUME

GENERAL CENSUS OF THE POPULATION OF 1962. RESULTS OF A 5 PERCENT SAMPLE; POPULATION, HOUSEHOLDS, INDIVIDUAL DWELLINGS, APARTMENT BUILDINGS; REGIONAL FASCICLES; SUMMARY FOR ALL OF FRANCE.
[22 fascicles; see listing on pp. 57-58 above.]

Structure of the Fascicles

There are three categories of table titles in the regional fascicles which are labeled as follows:

Number	Name	Table Titles Numbers
I.	Regional Tables	R 1 through R 23
II.	Departmental Tables	D 1 through D 17
III.	Agglomeration Tables	A 1 through A 9

Each regional fascicle contains data for all three categories of table titles, with the exception of the following:

Number	Region	Category Omitted
F.E.	All France	II, III
7.1	Poitou-Charentes	III

TABLES[1]

Number	Title

I. Regional Tables

R 1. Synoptic table of the principal information obtained from the 5 percent sample and the total enumeration.

R 2. Total population by sex, age, and marital status.

R 3. Total population by sex, age, and nationality.[2]

R 4. Economically active population, agricultural and nonagricultural, by employment status, sex, and age.

R 4A. Total population and economically active population, agricultural and nonagricultural households, by sex and age.

R 5. Economically active population by detailed employment status and sex.

R 6. Employed economically active population by sex, industrial division and major group, and employment status.

R 7. Economically active population by sex, socio-economic category, and employment status.

R 7A. Inactive population 15 years of age and over by socio-economic category.

---

[1] Page numbers not provided in original text.

[2] Native Frenchmen, naturalized citizens, foreigners, and Algerian Muslims.

FRANCE

R 8. Population of private households by the socio-economic category of the head of the household.

R 9. Pupils or students and apprentices 15 years of age and over by sex.

R 10. Pupils or students 11 years of age and over by instruction received.[3]

R 11. Non-student population (apprentices excluded) by sex, age, and educational level (general education up to high school; professional and technical education).

R 12. Inter-regional migrations from January 1, 1954, to March 7, 1962, by sex and age of the migrants. People living outside the metropolis on January 1, 1954, by sex and age.

R 13. Inter-regional migrations from January 1, 1954, to March 7, 1962, by place of origin and destination of the migrants.

R 14. Analysis of the composition of private households by category of commune.

R 15. Private households by number of persons, by the sex, marital status, and age of the head of the household.

R 16. Private households by number of persons, by the socio-economic category of the head of the household.

R 17. Principal characteristics of dwellings and their occupancy by category of commune.

R 18. Distribution of dwellings by the date they were built and by category of commune.

R 19. Distribution of residences by the number of rooms and the number of persons living in each, by category of commune.

R 20. Distribution of dwellings by the number of rooms, by category of commune.

R 21. Characteristics and occupancy status of housing by category of commune.

R 21A. Combined characteristics of dwellings by category of dwelling and category of commune.

---

[3]Categories of instruction received: (1) Elementary (primaire); (2) Short general (court général); (3) Long general (long général); (4) Short professional (court professionnel); (5) Long technical or professional (long technique ou professionnelle); (6) Agricultural (agricole); (7) Other technical or professional (autres techniques ou professionnelles); (8) Higher education (supérieur); and (9) Undetermined (indéterminé).

R 22. Multiple household structures by type, for each category of commune.

R 23. Characteristics of multiple household structures having at least one dwelling, by category of commune.

II. Departmental Tables

D 1. Synoptic tables of information obtained from the 5 percent sample and from the total enumeration.

D 2. Total population by sex, age, and marital status. Economically active population by sex and age.

D 3. Total population by sex and nationality.

D 4. Total population and economically active population, agricultural and nonagricultural households, by sex and age.

D 5. Economically active population by sex and detailed employment status.

D 6. Employed economically active population by sex, industrial division and major group, and employment status.

D 7. Economically active population by sex, socio-economic category, and employment status.

D 8. Population of private households by the socio-economic category of the head of the household.

D 9. Inter-departmental migrations from January 1, 1954, to March 7, 1962, and persons who lived outside the metropolis on January 1, 1954, by sex.

D 10. Private households by number of persons, by the socio-economic category of the head of the household.

D 11. Principal characteristics of dwellings and their occupancy status, by category of commune.

D 12. Distribution of dwellings by the date they were built, by category of commune.

D 13. Distribution of residences by number of rooms and number of occupants, by category of commune.

D 14. Distribution of dwellings by number of rooms, by category of commune.

D 15. Characteristics and occupancy status of residences by category of commune

D 15A. Combined characteristics of dwellings by category of dwelling and commune.

D 16. Multiple household structure by type, by category of commune.

D 17. Characteristics of multiple household structures which have at least one dwelling by category of commune.

FRANCE

III. Tables for Agglomerations

A 1. Synoptic table of principal information obtained by the 5 percent sample and by the total enumeration.

A 2. Total population by sex, age, and marital status. Economically active population by sex and age.

A 3. Economically active population by sex, socio-economic category, and employment status.

A 4. Private households by number of persons, by the socio-economic category of the head of the household.

A 5. Principal characteristics of dwellings and their occupancy.

A 6. Distribution of dwellings by date they were constructed.

A 7. Distribution of residences by number of rooms and number of occupants.

A 8. Characteristics and occupancy status of residences.

A 8A. Combined characteristics of dwellings by dwelling category.

A 9. Number of multiple household structures. Characteristics of multiple household structures which have at least one dwelling.

CENSUS OF 1962: POPULATION OF FRANCE; DEPARTMENTS, DISTRICTS, CANTONS, AND COMMUNES.

Structure of the Volume

This volume consists of three tables. Table I is presented only once at the beginning of the volume, but Tables II and III are repeated for each of the 90 departments consecutively in alphabetical order. A table of contents for the volume is not included.

Table I gives the total population (1962 definition) of each of the 90 departments and lists the number of districts, cantons, and communes in each department.

Table II gives tabulations for the cantons within each district of each department as follows:

Total population (1954 definition)
   1936
   1954
   1962

Total communal population (with the double count)

Municipal population (without the double count)

Total population (without the double count in the population enumerated separately)

Table III gives tabulations for each commune within each department, as follows:

Total population (1954 definition)
  1936
  1954
  1962

Total population (1962 definition)
  Municipal population
    Total
    In the administrative seat of the commune
    Outside the administrative seat
  Population enumerated separately
    Total
    Double count (those who belong to the municipal population of other communes)
    Total population without the double count

CENSUS OF 1962: CITIES OF MORE THAN 5,000 INHABITANTS.

Structure of the Volume

This volume is a compilation of tables, all of the same format, which present population figures for each city of more than 5,000 inhabitants. The tables are arranged in order of decreasing size of city population. The sets of population figures consist of:

Total population (1954 definition)
  1936
  1954
  1962

Ordinal number of the size of the city

Code number of the department in which the city is located

Name of the city

Total population (1962 definition)
  Municipal population
    Total
    In the administrative seat of the commune
    Outside the administrative seat
  Population enumerated separately
    Total
    Double count (those who belong to the municipal population of other communes)
    Total population without the double count

FRANCE

CENSUS OF 1962:   CITIES AND URBAN AGGLOMERATIONS.

TABLES

Number[1]	Title	Page
V.	Distribution by department of urban units[2] according to size of population.	14
VI.	List of urban units[2] of more than 20,000 people classified in decreasing order of magnitude.	18
VII.	List and composition of urban population clusters and detached towns by department.	23
VIII.	Summary table by category of population cluster.[3]	214

CENSUS OF 1962:   DE JURE POPULATION; STATISTICAL RESULTS.

TABLES

Number	Title	Page
P L 1.	De jure population, municipal population, and population enumerated separately, dwellings, mobile dwellings, and institutional households, by department in 1962.	18
P L 2.	Total population at different censuses; area and density in 1962.	
	Part 1. Departments. Population at successive censuses from 1801 to 1962.	22
	Part 2. Districts and departments. Total population in 1876, 1911, 1921, 1936, 1954, and 1962. Area and density in 1962.	26
	Part 3. Urban agglomerations and separate towns with more than 50,000 inhabitants in 1962. Total population in 1876, 1911, 1936, 1954, and 1962. Area and density in 1962.	36

---

[1] Tables I-IV consist of text tables and explanatory material.

[2] Urban units (unités urbaines): multicommunal agglomerations and separate towns. If an agglomeration extends over more than one department, it is counted as being in the department where the principal town or city of the agglomeration is located.

[3] Categories of population clusters: towns and urban agglomerations, rural communes (1962), and rural communes (1954).

TABLE TITLES BY VOLUME

        Part 4. Communes of more than 50,000 inhabitants in 1962. Total population in 1876, 1911, 1936, 1954, and 1962. Area and density in 1962.   40

P L 3. Comparisons with the 1936 and 1954 censuses. Variation between 1954 and 1962 discussed in terms of natural increase and migratory balance.

        Part 1. Departments. Total urban and rural communes of each department.   44

        Part 2. Agglomerations and communes of more than 50,000 inhabitants.   56

        Part 3. All France. Urban agglomerations and rural communes classified by size of population. Rural communes by category.[1] Zones of urban or industrial population.   64

P L 4. Classification of communes by size of population.

        Part 1. Departments. Total population and number of communes whose total population (with double count) is within a given range.   66

        Part 2. Departments. Number and total population of communes whose total population (with double count) is less than a given range.   78

        Part 3. All France. Number and population of the communes whose population (whether total population with double count, the 1954 definition, or without double count in the population enumerated separately) is within a given range.   90

        Part 4. All France. Number and population of the communes whose population (whether total population with double count, the 1954 definition, or without double count in the population enumerated separately) is less than a given range.   91

        Part 5. All France. Retrospective table.   92

P L 5. Population enumerated separately.

        Part 1. All France. Population enumerated separately by category[2] in 1954 and 1962 (1954 and 1962 definitions).   95

---

[1] Categories of rural communes: classification by commune size-class using the total population 1962 definition, and classification by percentage of the population dependent on agriculture using the 1954 definition.

[2] Categories of persons enumerated separately: military, persons in sanatoria and convalescent homes, students, workers

FRANCE

       Part 2. Departments. Population enumerated separately by category² in 1962 (1962 definition).   96

       Part 3. Departments. Population enumerated separately by category² in 1954 and 1962 (1954 definition).   98

P L 6. Institutional households. Number and population by category³ and department in 1962.   102

P L 7. Individual housing by department in 1861, 1911, 1921, 1936, 1946, 1954, and 1962.   104

CENSUS OF 1962: DE JURE POPULATION AND SUPPLEMENTARY STATISTICS FOR THE COMMUNES.

Structure of the Volumes

These volumes are a compilation of tables, all of the same format, for the 90 departments. The tables consist of sets of population figures for five different administrative, regional, and size-class categories. The sets of population figures consist of:

Total population (1954 definition)
  1936
  1954
  1962

Total population (1962 definition)
  Municipal population
    Total
    In the administrative seat of the commune
    Outside the administrative seat
  Population enumerated separately
    Total
    Double count (those who belong to the municipal population of other communes)
    Total population without the double count
  Municipal population in 1954

In addition, data on the components of population change (births, deaths, migration) between 1954 and 1962 are given, as well as density figures for residences and by square kilometers (1962).

---

lodged in temporary camps, prisoners, persons in psychiatric hospitals, and unknown.

³Categories of institutional households: religious institutions, old age homes or retreats, and others.

TABLE TITLES BY VOLUME

The five different administrative, regional, and population-cluster categories for which the population data are presented are:

1. Communes classed by canton and district
2. Towns and population clusters
3. Rural communes classed by the proportion of the population dependent on agriculture
4. Industrial and urban agglomerations
5. Agricultural regions

CENSUS OF 1962: INDUSTRIAL AND URBAN ZONES OF POPULATION.

Structure of the Volume

This volume is a compilation of tables, all of the same format, for the 90 departments. The tables consist of sets of population figures for industrial and urban zones of population. The sets of population figures consist of:

Total population (1954 definition)
  1936
  1954
  1962

Total population (1962 definition)
  Municipal population
    Total
    In the administrative seat of the commune
    Outside the administrative seat
  Population enumerated separately
    Total
    Double count (those who belong to the municipal population of other communes)
    Total population without the double count
  Municipal population in 1954

In addition, data on the components of population change (births, deaths, and migration) between 1954 and 1962 are given, as well as density figures for residences and by square kilometers (1962).

GERMANY

MAIN ENTRY AND VOLUME NOTES:

Germany (Federal Republic, 1949-   ) Statistisches Bundesamt.
  Bevölkerung und Kultur: Volks- und Berufszählung vom 6.
Juni 1961. Stuttgart, 1962-
    v. 30cm.   (79-363277)

Heft 1. Die methodischen Grundlagen der Volks- und Berufszählung
        1961. [1967] 267 p. [Excluded; no tables.]

Heft 2. Ausgewählte Bevölkerungsgruppen: Deutsche Bevölkerung
        und Ausländer. [1967] 122 p.

Heft 3. Bevölkerungsstand und Bevölkerungsentwicklung. [1966]
        144 p.

Heft 4. Bevölkerung nach Alter und Familienstand. [1966] 274 p.

Heft 5. Bevölkerung nach der Religionszugehörigkeit. [1966]
        92 p.

Heft 6. Vertriebene und Deutsche aus der SBZ: Verteilung und
        Struktur. [1967] 354 p.

Heft 7. Ausländer: Verteilung und Struktur. [1966] 232 p.

Heft 8. Bevölkerung in Anstalten. [1967] 98 p.

Heft 9. Pendler. [1967] 282 p.

Heft 10. Bevölkerung nach Lebensunterhalt und Beteiligung am
         Erwerbsleben. [1966] 111 p.

Heft 11. Bevölkerung und Erwerbspersonen mit überwiegendem
         Lebensunterhalt durch Angehörige bzw. Rente u. dgl.
         [1967] 117 p.

Heft 12. Erwerbspersonen in wirtschaftlicher und sozialer
         Gliederung. [1967] 195 p.

Heft 13. Erwerbspersonen in beruflicher Gliederung. [1968]
         376 p.

Heft 14. Erwerbstätige nach Wochenarbeitszeit und weiterer
         Tätigkeit. [1967] 91 p.

Heft 15. Personen mit einer abgeschlossenen Ausbildung. [1968]
         329 p.

Heft 16. Demographische und wirtschaftliche Struktur der
         Haushalte und Familien. (10% Aufbereitung) [1968]
         215 p.

Heft 17. Erwerbstätigkeit von Frauen und Müttern. (10% Aufbereitung) [1968] 115 p.

Heft 18. Kinder und Jugendliche in Familien. (10% Aufbereitung) [1967] 93 p.

Heft 19. Lebensverhältnisse der älteren Mitbürger. (10% Aufbereitung) [1967] 63 p.

Heft 20. Religionszugehörigkeit in Familien. (10% Aufbereitung) [1967] 80 p.

Heft 21. Untersuchungen zur Methode und Genauigkeit der Volks- und Berufszählung 1961. [Announced but not yet published.]

ENGLISH TRANSLATION OF CENSUS AND VOLUME TITLES:

Population and occupation census, June 6, 1961.

1. Methodological bases for the population and occupation census of 1961.
2. Selected population groups: German population and foreigners.
3. Status and growth of the population.
4. Population by age and marital status.
5. Population by religious affiliation.
6. Population of expellees and Germans from the SOZ[1]: distribution and structure.
7. Foreign population: distribution and structure.
8. Population in institutions.
9. Commuters.
10. Population by source of livelihood and participation in the labor force.
11. Population and economically active persons with principal source of livelihood from relatives, annuities, etc.
12. Industrial and social classification of economically active persons.
13. Occupational classification of economically active persons.
14. Employed persons by weekly work hours and additional activity.
15. Persons with completed education.

---

[1] Soviet Occupied Zone.

GERMANY

16. Demographic and industrial structure of households and families.[2]
17. Economic activity of married women and mothers.[2]
18. Children and youths in families.[2]
19. Principal source of livelihood of older citizens.[2]
20. Religious affiliation of families.[2]
21. Examination of the method and the exactitude of the population and occupation census of 1961. [Announced but not published.]

ENUMERATION DATE: June 6, 1961.

ENUMERATION BASIS: De jure.

GLOSSARY

ADMINISTRATIVE, GEOGRAPHICAL, AND POLITICAL UNITS

Level	English Translation	Vernacular Term
National	Federal Republic	Bundesgebiet
Intermediate	State	Land
Local	Self-governing district	Kreis
	Governmental district or Administrative district	Regierungsbezirk or Verwaltungsbezirk
	Commune	Gemeinde
Other concepts:		
	City	Stadt
	Independent city	Kreisfreie Stadt
	Rural district	Landkreis
	Commune size class	Gemeindegrosseklasse

States are the largest governmental units below the national level. They are divided into local self-governing districts, called either governmental districts or administrative districts, depending on the state. These districts are in turn divided into communes.

Communes and most cities are under the jurisdiction of the local self-governing district. Communes range in geographical and population size from rural villages with their surrounding farmlands to large cities. Large communes usually have the legal title of city. If the number of inhabitants in a city reaches a prescribed minimum, the city may be designated an independent

---
[2] 10 percent sample.

city and thereby become independent of the jurisdiction of the local self-governing district. **Rural districts** are the areas surrounding independent cities. (The word "rural" is not always appropriate, since many such districts include densely populated urban communes.) Rural districts correspond to American counties or English administrative counties, though their average size is smaller.*

In a number of tables, communes are grouped into **commune size-classes** as follows:

Communes with fewer than 2,000 inhabitants and 40 percent or more farm and forestry population;
Communes with fewer than 2,000 inhabitants and 20 to 40 percent farm and forestry population;
Communes with fewer than 2,000 inhabitants and less than 20 percent farm and forestry population;
Communes with 2,000 to 5,000 inhabitants;
Communes with 5,000 to 20,000 inhabitants;
Communes with 20,000 to 100,000 inhabitants;
Communes with 100,000 or more inhabitants (large cities).

TABLE OF DEMOGRAPHIC CONCEPTS

The demographic concepts which appear in the following table correspond to the recommendations of the European Conference of Statisticians, unless otherwise noted. They are defined and discussed in the appendixes and in footnotes. Terms in parentheses are alternative forms which appear in the table titles. Terms in brackets are not used in the table titles, but they are included here in order to give logical completeness to the various classifications in the table.

Appendix	English Table Term	Vernacular Table Term
	CULTURAL, EDUCATIONAL, AND PERSONAL CHARACTERISTICS	
...	Age[1]	Alter
	Commuters	Pendler[2]
	Occupational commuters[2]	Berufspendler
	Educational commuters	Ausbildungspendler
	Completed education[3]	Abgeschlossene Ausbildung
	Trade school	Berufsschule
	Vocational school	Berufsfachschule

---

*J.F.J. Gillen, State and Local Government in West Germany, 1945-1953 (Office of the U.S. High Commissioner for Germany, 1953), p. 2.

[1][All footnotes are at the end of the Table.]

Professional school	Fachschule
College/university	Hochschule
Religious affiliation[4]	Religionszugehörigkeit

EXPELLEES[5] AND GERMANS FROM THE SOVIET OCCUPIED ZONE

... [Expellees] [Vertriebene]
    German citizens, with their children, who claim to possess a federally issued Identity Card A or B   Ausweis-Vertriebene
    [Persons expelled from their homelands]   [Wohnsitz-Vertriebene]
    Germans from the Soviet Occupied Zone[6]   Deutsch aus der SBZ
    Refugees from the Soviet Zone   Sowjetzonenflüchtlinge

ECONOMIC CHARACTERISTICS

A   Economic activity (participation in labor force)   Stellung zum Erwerbsleben
    Economically active[7]   Erwerbspersonen
    Economically inactive   Nichterwerbspersonen
    Not employed[8]   Nichterwerbstätige
    Income recipients[9]   Einkommensbezieher

B   Employment status   Stellung im Beruf

C   Industrial classification[10]   Wirtschaftslicher Gliederung
    Division(s)   Wirtschaftsabteilungen
    Major group(s)   Wirtschaftsunterabteilungen
    Group(s)   Wirtschaftsgruppen

D   Occupational classification   Beruflicher Gliederung
    Major group(s)[11]   Berufsabteilungen
    Minor group(s)[11]   Berufsgruppen
    Unit group(s)   Berufsordnungen
    Unit subgroup(s) (Individual occupations)   Berufsklassen
    Additional occupation[12]   Weitere Tätigkeit
    Principal source of livelihood[13]   Überwiegender Lebensunterhalt

E   Socio-economic category (Social status)   Soziale Stellung

HOUSEHOLD AND FAMILY CHARACTERISTICS

F   [Household]   ...
    Private household[14]   Privathaushalt
    Head of household[15]   Haushaltsvorstand

GLOSSARY

Generations within a household[16]	Generationen
Types of household[17]	Haushaltstypen
Institution[18]	Anstalt
Family[19]	Familie
Type of family[20]	Familientyp
Head of family[21]	Familienvorstand
Marital status[22]	Familienstand

---

[1] The classification by age throughout most of the census is based on age at the last birthday. Exceptions occur in Volumes X, XVI, and XIX, where age is based on year of birth and is equivalent to the difference between that year and the census year (1961).

[2] For occupational commuters, the place of work (Arbeitsort) is the place where the work is actually performed, not the location of the firm or enterprise.

[3] The classification by completed education is based on the highest level of education for which a person has passed a final examination. There are four levels of education above the secondary school level. Trade schools are attended by persons employed in an occupation during a training period; attendance is compulsory. These schools, established shortly after World War I to replace the previous "continuation schools," are provided for youths under 18 years of age, whether or not they have completed earlier schooling. Instruction is offered one or two days weekly. Vocational schools, which prepare students for particular occupations, are attended voluntarily by youths under 18 years of age who have left earlier schooling. They offer at least a full year of full-time instruction. Examples are commercial schools, household and children's nursing schools, etc. Professional schools, which are voluntarily attended, provide education in an occupation beyond the trade school level for persons 18 years of age or older who qualify on the basis of earlier training. Full-time instruction (30 or 40 weekly lessons) lasting at least half a year leads to more advanced vocational standing. Examples are construction, maritime, and nursing schools. College/university corresponds to a college or university in the United States. The classification comprises Universitäten, Technische Hochschulen, Wirtschaftshochschulen, and Akademien. Excluded from the concept are Forschung and Volkshochschulen.

Determining the characteristics of people possessing what in the United States is called a "college degree" was considered particularly important in this census. Information on pupils and students in schools other than college level has in the past been analyzed in depth, but this census is the first to present detailed information about the population with a college education. Such data, in combination with data on economic characteristics, are useful in evaluating the occupational classifications of the generation just entering the labor market.

[4] The classification by religious affiliation is based on legal (rechtliche) membership in a denomination, not to confession of religious belief. In accordance with the List of Religious Denominations (1961), the eleven denominational categories are:

Churches combined into the Protestant Church in Germany
Protestant Free Churches (e.g., Lutheran Free Churches, Salvation Army)
Roman Catholic Church
Eastern Orthodox Churches
Old Catholic and related groups
Christian-oriented separate religions (e.g., Seventh Day Adventists)
Jewish religious community
Other world religions
Independent religious and ideological communities (e.g., Theosophists)
No denomination
Unclear or defective indication of religious belief

[5] Expellees in the 1961 census include those German citizens, with their children, who claim to possess a federally issued Identity Card A or B (Ausweis-Vertriebene). Such persons were residents on September 1, 1939, of those eastern German territories which belonged to Germany prior to World War II and passed to other countries (e.g., Poland) after the war. These persons, after their expulsion after World War II, either came directly into the Federal Republic or settled first in the Soviet Occupied Zone before emigrating to the Federal Republic.

Since all persons expelled from their homelands have not acquired federally issued Identity Cards A or B, the continuous registration of expellees will give larger totals than the enumeration of identity card holders.

For a detailed discussion of expellees in the Federal Republic of Germany, see Paul F. Meyer and W. Parker Mauldin, "Assimilation of the Expellees in the Federal Republic of Germany, Demographic and Social Aspects," in United Nations, Department of Social and Economic Affairs, Population Bulletin (ST/SOA/Series N/No. 2, October 1952), pp. 15-26.

[6] Germans from the Soviet Occupied Zone are German citizens who came from the Soviet Occupied Zone into the Federal Republic after World War II. These persons may or may not have been residents of the area which is now the Soviet Occupied Zone prior to Soviet occupation.

GLOSSARY

Refugees from the Soviet Zone are German citizens who fled to the Federal Republic from the area which is now the Soviet Occupied Zone. Such persons were residents of that area on September 1, 1939. They possess federally issued Identity Cards C and form a sub-category of Germans from the Soviet Occupied Zone.

[7] Economically active: Persons working for pay or profit (employed, Erwerbstätige, and unemployed, Erwerbslose). The status of the economically active population with reference to the labor market (that is, whether persons are employed or unemployed) is referred to as Beteiligung am Erwerbsleben (translated as type of participation in the labor force).

[8] In some tables, the unemployed and the economically inactive are grouped together as not employed, in contrast to the employed.

[9] Income recipients are persons classified as receiving or not receiving an income on the basis of their participation in the labor force (i.e., economic activity).

INCOME RECIPIENTS

Economic Activity Category	Subsistence Category: Principal Source of Livelihood			
	Employment	Unemployment Funds/Aid	Rents and Annuities	Relatives
Employed	Inc. Rec.	Inc. Rec.	Inc. Rec.	Inc. Rec.
Unemployed	X	Inc. Rec.	Inc. Rec.	NOT Inc. Rec.
Economically Inactive	X	X	Inc. Rec.	NOT Inc. Rec.

[10] In the table titles, "industrial branch" (Wirtschaftsbereich) is an abbreviated industrial classification designating tabulations of four categories:

1. Agriculture, forestry, fishing
2. Manufacturing
3. Commerce, transportation
4. Remaining industries (service)

[11] Tabulations of these groups are not given. The term appears only in the introduction to the census.

[12] An additional occupation is an auxiliary occupation no matter how minor, but it is registered only if it was engaged in at the

time of the census (an occupation learned but not exercised is not registered). Honorary activities such as juryman, guardian, etc., are not included.

[13] The principal source of livelihood was ascertained for each person. For persons who are predominantly supported by relatives (by, for example, parents or a husband), the principal source of livelihood of the breadwinner is indicated. Sources of livelihood are categorized as follows: (a) employment (Erwerbstätigkeit), (b) unemployment funds/aid (Arbeitslosengeld/--hilfe), (c) annuity, similar incomes (Rente), (d) relatives (Angehörige).

[14] In the table titles, "private household" designates every community of persons living together and sharing responsibility for housekeeping. Private households may be composed of related or unrelated persons, including household personnel or agricultural workers. Institutions are not considered households, but a household can exist within the premises of an institution (e.g., the household of the manager of the institution). A person keeping house by himself is considered a private household.

[15] The head of the household is the person, 15 years of age or older, who claims to be the head of the household.

[16] Generations within a household are reckoned according to the relationship of individual household members to the head of the household.

[17] Types of household derive from the types of relationships between members of a household (persons unrelated, or related by affinal and generational ties) as follows:

Household Type	Composition of Household
A 1	Childless couple
A 2	Parents and single children or grandchildren
A 3	Parents and married children (without grandchildren) as well as possibly some unmarried children
A 4	Parents, children, grandparents, possibly grandchildren of the parents
B 1	Types A 1 through A 4 in which other relatives live who are related consanguineally or affinally in a direct line
B 2	Types A 1 through A 4 in which persons live who are not related either consanguineally or affinally in a direct line
C 1	Types A 1 through B 1, including persons alien to the family (such as servants), excluding Gaststätte establishments

GLOSSARY

C 2   Type B 2 including persons alien to the family (such as servants), excluding Gaststätte establishments

D   Unrelated persons having a joint household (e.g., two retired school teachers), excluding Gaststätte establishments

G   Types C 1, C 2, and D where the owner of a Gaststätte establishment lives, having at least one but no more than five servants

[18] Institutions are public or private organizations which serve a specific social, religious, health, educational, or other purpose in the lodging, feeding, and care of people. Large accommodation establishments and hotels are enumerated as institutions. The institutional population includes resident staff members and inmates, if these do not belong to private households within the institutions or are not temporary staff members or inmates.

[19] The family is a group of related persons living together in the same household. Families are classified as follows:

Parents with their unmarried children
Single parents with their unmarried children
Childless couples
Couples whose children no longer live in the household
Widowed or divorced persons not living with their children (one-person families)

A single person residing apart from his parents or his own children is not considered a family.

[20] Type of family is derived from a combination of marital status and presence or absence of own children. From this cross-classification, types F 1 through F 8 emerge. F 3 is the sole type that does not relate to own children. It is "Married couples with grandchildren only." The derivation of types of family from marital status and presence or absence of own children is shown below.

Marital Status	Children Present	
	Yes	No
Single	F 6	...
Married, living with spouse	F 2	F 1
Married, spouse missing/separated	F 8	F 7
Widowed	F 5	F 4
Divorced (annulled)	F 5	F 4

Children are included regardless of age. Stepchildren and adopted children are counted as children. Foster children are not counted as children but as persons related to the family.

[21] The head of the family is, in the case of a couple, the husband. If a married woman did not provide information about her spouse, she is considered the head of the family. Widowed and divorced persons, with or without children, as well as single persons living with their children, are enumerated as family heads.

[22] The marital states of the population fall into the categories of single, married, widowed, and separated. Persons whose spouses are missing but not declared dead are regarded as married. Married couples are regarded as separated if husband and wife belong to the resident population of different communes. Year of marriage refers to calendar year; persons married between January 1, 1961, and the census date were considered married in 1961.

TABLE TITLES BY VOLUME

VOLUME 2. SELECTED POPULATION GROUPS: GERMAN POPULATION AND FOREIGNERS.

TABLES

Number	Title	Page
1.	Germans and foreigners by independent city and rural district.	40
2.	Germans and foreigners by age group, marital status, and state.	77
3.	Germans and foreigners by year-of-birth group and state.	101
4.	Germans and foreigners by religious affiliation and state.	105
5.	Economically active Germans and foreigners by age group and age-specific activity rate in the Federal Republic.	106
6.	Economically active Germans and foreigners by industrial subdivision and employment status in the Federal Republic.	107
7.	Economically active Germans and foreigners by industrial division and employment status in the Federal Republic.	109
8.	Germans and foreigners by participation in the labor force in the Federal Republic.	110
9.	Germans and foreigners whose education was completed at a vocational or professional school by age group in the Federal Republic.	111

10.	Germans and foreigners whose education was completed at a college/university by age group in the Federal Republic.	112
11.	Germans and foreigners whose education was completed at a vocational or professional school by major subject in the Federal Republic.	113
12.	Germans and foreigners whose education was completed at a college/university by major subject in the Federal Republic.	114
13.	Germans and foreigners whose education was completed at a college/university by year-of-completion group and major subject in the Federal Republic.	115
14.	Germans and foreigners whose education was completed at a college/university by year-of-completion group in the Federal Republic.	116
15.	Families by family type and expellee/refugee status of the head of the family in the Federal Republic.	116
16.	Private households of Germans and foreigners by state.	117
17.	Households by participation in the labor force, expellee/refugee status of the head of the household, and household type in the Federal Republic.	118

VOLUME 3. STATUS AND GROWTH OF THE POPULATION.

TABLES

Number	Title	Page
1.	Area, resident population, population density, and population growth in independent cities and rural districts, 1939, 1950, and 1961.	52
2.	Resident population of communes by commune size-class in the districts on 6-6-1961.	88
3.	Population growth by commune size-class in the governmental districts, 1939-1950, 1939-1961, and 1950-1961.	110
4.	Resident population of communes and persons with additional living quarters on 6-6-1961 by commune size-class in the states.[1]	116

---

[1] Persons with additional living quarters (Weiterer Wohnraum) are members of a household who maintain another apartment or room for work or study, separate from their usual place of residence.

GERMANY

5.	Resident population of communes by commune size-class in the states.	118
6.	Resident population of communes and the proportion in agriculture or forestry by commune size-class in governmental and administrative districts and states.	120

VOLUME 4. POPULATION BY AGE AND MARITAL STATUS.

TABLES

Number	Title	Page
1.	Resident population by year of birth and marital status in the Federal Republic.	58
2.	Population growth by year of birth and sex, 1950-1961, in the Federal Republic.	62
3.	Resident population by age and marital status in the Federal Republic.	66
4.	Resident population by age and marital status in the states.	70
5.	Resident population by age and marital status, 1950 and 1961, in the Federal Republic.	114
6.	Resident population by age group and marital status in independent cities and rural districts.	120
7.	Resident population by age group and commune size-class in the states.	244

VOLUME 5. POPULATION BY RELIGIOUS AFFILIATION.

TABLES

Number	Title	Page
1.	Resident population by age group and religious affiliation in the states.	38
2.	Resident population by religious affiliation in independent cities and rural districts.	62
3.	Resident population by commune size-class and religious affiliation in the states.	84

TABLE TITLES BY VOLUME

VOLUME 6. POPULATION OF EXPELLEES AND GERMANS FROM THE SOVIET OCCUPIED ZONE: DISTRIBUTION AND STRUCTURE.

TABLES

Number	Title	Page
1.	Expellees and Germans from the Soviet Occupied Zone by religious affiliation in the states.	65
2.	Expellees and Germans from the Soviet Occupied Zone by independent city and rural district.	66
3.	Expellees and Germans from the Soviet Occupied Zone by commune size-class in the states.	78
4.	Expellees and Germans from the Soviet Occupied Zone by commune size-class in the states.	80
5.	Expellees with Identity Card A or B by age and marital status in the Federal Republic.	82
6.	Germans from the Soviet Occupied Zone without Identity Card A or B by age and marital status in the Federal Republic.	86
7.	Refugees from the Soviet Occupied Zone with Identity Card C by age and marital status in the Federal Republic.	90
8.	Expellees with Identity Card A or B by age, marital status, and type of participation in the labor force, and economically inactive persons by principal source of livelihood, in the Federal Republic.	94
9.	Germans from the Soviet Occupied Zone without Identity Card A or B by age, marital status, and type of participation in the labor force, and economically inactive persons by principal source of livelihood, in the Federal Republic.	96
10.	Refugees from the Soviet Occupied Zone with Identity Card C by age, marital status, and type of participation in the labor force, and economically inactive persons by principal source of livelihood, in the Federal Republic.	98
11.	Economically active expellees and Germans from the Soviet Occupied Zone by industrial division and social status in the Federal Republic.	100
12.	Expellees with Identity Card A or B by type of participation in the labor force, age group, marital status, and principal source of livelihood in the Federal Republic.	110

13. Germans from the Soviet Occupied Zone without Identity Card A or B by type of participation in the labor force, age group, marital status, and principal source of livelihood in the Federal Republic. 116

14. Refugees from the Soviet Occupied Zone with Identity Card C by type of participation in the labor force, age group, marital status, and principal source of livelihood in the Federal Republic. 122

15. Expellees with Identity Card A or B whose principal source of livelihood is from relatives by participation in the labor force, industrial division, and employment status of the breadwinner in the Federal Republic. 128

16. Germans from the Soviet Occupied Zone without Identity Card A or B whose principal source of livelihood is from relatives by participation in the labor force, industrial division, and employment status of the breadwinner in the Federal Republic. 134

17. Economically active expellees with Identity Card A or B whose principal source of livelihood is from relatives by participation in the labor force, industrial division, and employment status of the breadwinner in the Federal Republic. 140

18. Economically active Germans from the Soviet Occupied Zone without Identity Card A or B whose principal source of livelihood is from relatives by participation in the labor force, industrial division, and employment status of the breadwinner in the Federal Republic. 146

19. Expellees with Identity Card A or B whose principal source of livelihood is from relatives by age group and by participation in the labor force, industrial subdivision, and employment status of the breadwinner in the Federal Republic. 152

20. Germans from the Soviet Occupied Zone without Identity Card A or B whose principal source of livelihood is from relatives by age group and by participation in the labor force, industrial subdivision, and employment status of the breadwinner in the Federal Republic. 162

21. Expellees with Identity Card A or B whose education was completed at a vocational school, professional school, or college/university by major subject of education, participation in the labor force, and age group in the Federal Republic. 172

22.	Germans from the Soviet Occupied Zone without Identity Card A or B whose education was completed at a vocational school, professional school, or college/university by major subject of education, participation in the labor force, and age group in the Federal Republic.	188
23.	Expellees with Identity Card A or B in institutions by type of institution, age group, staff or inmate status, and marital status in the Federal Republic.	204
24.	Germans from the Soviet Occupied Zone without Identity Card A or B in institutions by type of institution, age group, staff or inmate status, and marital status in the Federal Republic.	207
25.	Economically active expellees with Identity Card A or B in institutions by type of institution, staff or inmate status, industrial branch, age group, principal source of livelihood, and marital status in the Federal Republic.	210
26.	Economically active Germans from the Soviet Occupied Zone without Identity Card A or B in institutions by type of institution, staff or inmate status, industrial branch, age group, principal source of livelihood, and marital status in the Federal Republic.	214
27.	Economically inactive expellees with Identity Card A or B who are inmates in institutions by type of institution, age group, principal source of livelihood, and marital status in the Federal Republic.	218
28.	Economically inactive Germans from the Soviet Occupied Zone without Identity Card A or B who are inmates in institutions by type of institution, age group, principal source of livelihood, and marital status in the Federal Republic.	221
29.	Multi-person households by source of income of income recipients, type of household, and number of income recipients.	224
30.	Multi-person households by type of household and by participation in the labor force, industrial branch, employment status, and age of the head of the household.	228
31.	Selected types of multi-person households by number of children and by marital status, participation in the labor force, industrial branch, and employment status of the head of the household.	252
32.	Persons in single person households by marital status, principal source of livelihood, industrial branch, employment status, sex, and age.	264

33.	Married couples by number of children, year-of-marriage group, and religious affiliation of spouses.	272
34.	Selected types of families by source of income of income recipients and number of income recipients.	290
35.	Selected types of families by number of children and by participation in the labor force, industrial branch, and employment status of the head of the family.	292
36.	Married couples with children by number and age of children, type of participation in the labor force and employment status of spouses, and by industrial branch of the husband.	300
37.	Unmarried children in families, 15 years of age and older, by participation in the labor force and employment status of the children and the head of the family, and by family type and number of children in the family.	306
38.	Outgoing and incoming civilian commuters (occupational commuters, pupils, and students) who are expellees with Identity Card A or B by independent city and rural district.	312
39.	Outgoing and incoming civilian commuters (occupational commuters, pupils, and students) who are Germans from the Soviet Occupied Zone without Identity Card A or B by independent city and rural district.	327

VOLUME 7. FOREIGN POPULATION: DISTRIBUTION AND STRUCTURE.

TABLES

Number	Title	Page
1.	Foreigners by citizenship in the states.	34
2.	Foreigners and economically active foreigners by age in the states.	38
3.	Foreigners by age group and citizenship in the states.	54
4.	Foreigners by age group, marital status, and citizenship in the Federal Republic.	78
5.	Foreign pupils and students by age in the states.	84
6.	Foreign pupils and students by citizenship in the states.	86

7.	Foreigners by religious affiliation[1] in governmental and administrative districts, by state.	90
8.	Foreigners in private households and institutions by citizenship in the states.	92
9.	Economically active foreigners by selected individual occupation, employment status, principal source of livelihood, and citizenship in the Federal Republic.	98
10.	Foreigners by type of participation in the labor force and citizenship in the Federal Republic.	152
11.	Economically active foreigners by principal source of livelihood, industrial division, and employment status in the Federal Republic.	156
12.	Economically active foreigners by citizenship and principal source of livelihood in the Federal Republic.	158
13.	Economically active foreigners by age group and weekly work hours in the Federal Republic.	159
14.	Economically inactive foreigners by age group, marital status, and principal source of livelihood in the Federal Republic.	160
15.	Foreigners with principal source of livelihood from relatives by participation in the labor force, industrial division, and employment status of the breadwinner in the Federal Republic.	162
16.	Economically active foreigners with completed education by occupational unit group, major subject group of education, employment status, and age group in the Federal Republic.	164

---

[1] Religious affiliation refers to membership in a denomination, not to confession of religious belief. Foreigners have been classified by religious affiliation in the following categories:

Protestants
Roman Catholics
Orthodox Catholics
Persons belonging to "other" denominations (Jewish, etc.)
Persons not belonging to any religious denomination.

GERMANY

VOLUME 8. POPULATION IN INSTITUTIONS.

TABLES

Number	Title	Page
1.	Institutional population by type of institution, age group, staff or inmate status, and marital status in the Federal Republic.	32
2.	Economically active persons among institutional population by type of institution, staff, inmates, industrial branch, age group, principal source of livelihood, and marital status in the Federal Republic.	44
3.	Economically inactive inmates in institutions by type of institution, age group, principal source of livelihood, and marital status in the Federal Republic.	80

VOLUME 9. COMMUTERS.

TABLES

Number	Title	Page
1.	Occupational commuters by district: June 6, 1961.	46
2.	Economically active persons and pupils and students (who are economically inactive) by commune size-class of place of work, school, or study, and economically active persons by industrial division in the Federal Republic.[1]	62
3.	Outgoing commuters among economically active persons and among pupils and students (who are economically inactive) by place of residence, and employed outgoing commuters by industrial division, in independent cities, rural districts, and communes belonging to a district with 10,000 or more inhabitants, by state.	64
4.	Incoming commuters among economically active persons and among pupils and students (who are economically inactive) by place of work, school, or study, and employed incoming commuters by industrial	

---

[1] A part of the student population is working or seeking work for gain, and hence is included as part of the economically active population. The remaining students are economically inactive.

division, in independent cities, rural districts, and communes belonging to a district with 10,000 or more inhabitants, by state. 164

5. Outgoing commuters among economically active persons and among pupils and students (who are economically inactive) by place of residence, and employed commuters by industrial division, by type of commune^2 in the states. 264

6. Incoming commuters among economically active persons and among pupils and students (who are economically inactive) by place of work, school, or study, and employed incoming commuters by industrial division, by type of commune^2 in the states. 272

VOLUME 10. POPULATION BY SOURCE OF LIVELIHOOD AND PARTICIPATION IN THE LABOR FORCE.

TABLES

Number	Title	Page
1.	Resident population by type of participation in the labor force, age group, marital status, and principal source of livelihood.	38
2.	Resident population by age, marital status, and participation in the labor force.	64
3.	Resident population by commune size-class, principal source of livelihood, and industrial division of the breadwinner.	100
4.	Breadwinners whose principal source of livelihood is from employment, from unemployment funds/aid, or from relatives by commune size-class and employment status.	102

VOLUME 11. POPULATION AND ECONOMICALLY ACTIVE PERSONS WITH PRINCIPAL SOURCE OF LIVELIHOOD FROM RELATIVES, ANNUITIES, ETC.

TABLES

Number	Title	Page
1.	Persons whose principal source of livelihood is from relatives by participation in the labor force,	

---

^{2}Type of commune refers to the commune size-class and, for

	industrial division, and employment status of the breadwinner in the Federal Republic.	26
2.	Economically active persons whose principal source of livelihood is from relatives by participation in the labor force, industrial division, and employment status of the breadwinner in the Federal Republic.	40
3.	Persons whose principal source of livelihood is from relatives by age group and by participation in the labor force, industrial major group, and employment status of the breadwinner in the Federal Republic.	54
4.	Economically active persons whose principal source of livelihood is from annuity and similar incomes or from relatives by age group, marital status, and employment status in the Federal Republic.	86
5.	Resident population whose principal source of livelihood is from annuity and similar incomes or from relatives by type of participation in the labor force, age group, and marital status in the Federal Republic.	96

VOLUME 12. INDUSTRIAL AND SOCIAL CLASSIFICATION OF ECONOMICALLY ACTIVE PERSONS.

TABLES

Number	Title	Page
1.	Economically active persons by age, woman's marital status, type of participation in the labor force, and employment status.	42
2.	Economically active persons by industrial division, employment status, and type of participation in the labor force.	58
3.	Economically active persons by commune size-class and employment status.	79
4.	Economically active persons by industrial group, employment status, age group, and principal source of livelihood.	98

---

communes with fewer than 2,000 inhabitants, the proportion of the population in agriculture and forestry. (See Glossary: Administrative, Geographical, and Political Units.)

TABLE TITLES BY VOLUME

VOLUME 13. OCCUPATIONAL CLASSIFICATION OF ECONOMICALLY ACTIVE PERSONS.

TABLES

Number	Title	Page
1.	Economically active persons by individual occupation, employment status, age group, and principal source of livelihood in the Federal Republic.	54
2.	Economically active persons by individual occupation in the states.	256
3.	Economically active persons by individual occupation, employment status, and industrial group in the Federal Republic.	290
4.	Diagrams: Economically active persons by selected occupational unit group, individual occupation, age group, and sex in the Federal Republic.	331

VOLUME 14. EMPLOYED PERSONS BY WEEKLY WORK HOURS AND ADDITIONAL ACTIVITY.

TABLES

Number	Title	Page
1.	Employed persons by industrial division, employment status, age group, and weekly work hours.	30
2.	Employed persons by age group, marital status, and additional activity.	54
3.	Employed persons by industrial division, employment status, and additional activity.	58
4.	Employed persons by occupational unit group, employment status, and additional activity.	63

VOLUME 15. PERSONS WITH A COMPLETED EDUCATION.

TABLES

Number	Title	Page
1.	Persons whose education was completed at a vocational school, professional school, or college or university by major field of study, source of income, and age group.	36

GERMANY

    (a) Graduates of vocational schools and professional schools.   36
    (b) Graduates of colleges and universities.   108

2. Economically active persons whose education was completed at a vocational school, professional school, or college or university by occupational minor group, major field of study, and [five-year] age group.   200
    (a) Graduates of vocational schools and professional schools.   200
    (b) Graduates of colleges and universities.   268

VOLUME 16. DEMOGRAPHIC AND INDUSTRIAL STRUCTURE OF HOUSEHOLDS AND FAMILIES. (10 PERCENT SAMPLE.)

TABLES

Number	Title	Page
1.	Multi-person households by the number of generations in the household. Members of multi-person households by generation sequence, type of participation in the labor force, and relationship to the head of the household.	47
2.	Members of multi-person households by number of generations in the household, household type and size, marital status, and age.	50
3.	Members of multi-person households by household size, marital status, age, and commune size-class.	54
4.	Multi-person households by number of generations in the household, household type and size, and marital status and age of the head of the household.	56
5.	Multi-person households and their members by number of generations in the household and household type and size. Members of multi-person households by sex, marital status, and age.	58
6.	Heads of households and members of multi-person households by number of generations in the household, household type and size, and type of participation in the labor force.	60
7.	Multi-person households by number of income recipients and household type and size.	62
8.	Selected types of multi-person households by number of income recipients, number of children, and household size.	64

		Page
	(a) Total.	64
	(b) In communes with 100,000 or more inhabitants.	66
9.	Multi-person households by source of income of income recipients, household type, and number of income recipients.	68
10.	Multi-person households by participation in the labor force, industrial branch, employment status, and age of the head of the household and by household type.	76
11.	Selected types of multi-person households by marital status, participation in the labor force, industrial branch, and employment status of the head of the household and by number of children.	84
12.	Multi-person households with income recipients by number of income recipients and their participation in the labor force, and by industrial branch, employment status, and age of the head of the household.	88
	(a) All multi-person households.	88
	(b) Multi-person households excluding types C 1 and G.	90
13.	Multi-person households by household type and by participation in the labor force, industrial branch, employment status, age, and expellee/refugee status of the head of the household.	92
14.	Selected types of multi-person households by marital status, participation in the labor force, industrial branch, employment status, and expellee/refugee status of the head of the household and by number of children.	100
15.	Multi-person households with income recipients by number of income recipients and by participation in the labor force, industrial branch, employment status, age, and expellee/refugee status of the head of the household.	108
16.	Multi-person households with one or more income recipients (other than head of household) by age, participation in the labor force, principal source of livelihood, and expellee/refugee status of the head of the household and by number of additional income recipients and their principal source of livelihood.	112
17.	Selected types of families by number of income recipients and number of children.	120
18.	Selected types of families by number of income recipients and number of children who are income recipients.	124

19.	Selected types of families by source of income of income recipients and number of income recipients.	128
20.	Selected types of families by participation in the labor force, industrial branch, employment status, and age of the head of the family.	132
21.	Selected types of families by participation in the labor force, industrial branch, and employment status of the head of the family and by number of children.	136
22.	Selected types of families with income recipients by participation in the labor force, industrial branch, employment status, and age of the head of the family and by number of income recipients.	140
23.	Families with one or more income recipients (other than the head of the family) by age, participation in the labor force, and principal source of livelihood of the head of the family and by number of additional income recipients and their principal source of livelihood.	144
24.	Selected types of families by number of income recipients, number of children who are income recipients, and expellee/refugee status of the head of the family.	150
25.	Selected types of families by participation in the labor force, industrial branch, employment status, and expellee/refugee status of the head of the family and by number of children.	158
26.	Multi-person households by number of income recipients, household type, state, and commune size-class.	166
27.	Multi-person households by types A 2 through A 4, B 1, C 1, and G by number of children and number of income recipients, by state.	168
28.	Multi-person households by participation in the labor force, industrial branch, employment status, and sex of the head of the household, by state.	170
29.	Households of type A 2 by participation in the labor force, industrial branch, employment status, and sex of the head of the household, by state.	172
30.	Multi-person households of types A 2 through A 4, B 1, C 1, and G by marital status and sex of the head of the household and household type, by state.	174
31.	Multi-person households of types A 2 through A 4, B 1, C 1, and G by participation in the labor force, industrial branch, and employment status of the head	

	of the household, household type, and number of children in the household, by state.	182
32.	Selected types of families by number of income recipients and number of children, by state.	190
33.	Selected types of families by number of children and number of children who are income recipients, by state.	198
34.	Families of types F 1, F 2, and F 5 by participation in the labor force, industrial branch, and employment status of the head of the family, by state.	202
35.	Families of type F 1 (couples without children) by participation in the labor force, industrial branch, and employment status of the head of the family, by state.	203
36.	Families of type F 2 (couples with children) by participation in the labor force, industrial branch, and employment status of the head of the family, by state.	204
37.	Families of type F 5 (with widow as head) by participation in the labor force, industrial branch, and employment status of the head of the family, by state.	205
38.	Selected types of families by participation in the labor force, industrial branch, and employment status of the head of the family and by number of children, by state.	206

VOLUME 17. ECONOMIC ACTIVITY OF MARRIED WOMEN AND MOTHERS. (10 PERCENT SAMPLE.)

TABLES

Number	Title	Page
	Results for the Federal Republic	
1.	Married women and female heads of families by age, participation in the labor force, and type of family.	58
2.	Married women and female heads of families by participation in the labor force, industrial branch, employment status, and type of family.	59
3.	Employed married women and female heads of families by age, employment status, and type of family.	60
4.	Employed married women and female heads of families by employment status, weekly work hours, and type of family.	61

GERMANY

5. Employed married women and female heads of families, with principal source of livelihood from employment, by employment status, weekly work hours, and type of family. — 62

6. Employed married women and female heads of families, who are employees, by weekly work hours, time spent daily traveling to work, and type of family. — 63

7. Mothers by number and age of children, type of family, and type of participation in the labor force. — 64

8. Mothers by type of participation in the labor force, age of children, and type of family.
    (a) By age. — 68
    (b) By commune size-class. — 70

9. Employed mothers by number and age of children, type of family, and employment status. — 72

10. Employed mothers, who are employees, by age of children, weekly work hours, time spent daily traveling to work, type of family, and industrial branch. — 76

11. Couples without children by participation in the labor force, industrial branch, and employment status of spouses. — 80

12. Couples with children by participation in the labor force, industrial branch, and employment status of spouses. — 80

13. Couples with children by number and age of children, industrial branch of the husband, and type of participation in the labor force and employment status of spouses. — 82

14. Couples with children by number of children, industrial branch of the husband, commune size-class, and type of participation in the labor force and employment status of spouses. — 88

15. Couples by number of children, industrial branch of the head of the family, and type of participation in the labor force, employment status, and religious affiliation of spouses. — 94

Results by States

16. Married women and female heads of families by participation in the labor force and type of family. — 96

17. Married women and female heads of families by age and participation in the labor force. — 97

18. Married women and female heads of families by participation in the labor force, industrial branch, and employment status. — 98

19.	Employed married women and female heads of families by age and employment status.	99
20.	Employed married women and female heads of families by employment status and weekly work hours.	100
21.	Employed married women and female heads of families, who are employees, by weekly work hours and time spent daily traveling to work.	101
22.	Mothers by number and age of children.	102
23.	Employed mothers by number and age of children.	103
24.	Married women without children (F 1) by age and participation in the labor force.	104
25.	Married women with children (F 2) by age and participation in the labor force.	105
26.	Employed married women with children (F 2) by number and age of children.	106
27.	Married women with children (F 2) who are self-employed or are family helpers by number and age of children.	107
28.	Employed married women with children (F 2) who are employees by number and age of children.	108
29.	Employed married women with children (F 2) by age of children and commune size-class.	109
30.	Married women with children (F 2) who are not employed by number and age of children.	110
31.	Married women with children (F 2) who are not employed by age of children and commune size-class.	111
32.	Married women without and with children (F 1 and F 2) by participation in the labor force, industrial branch, and employment status of husband.	112
33.	Employed married women without and with children (F 1 and F 2) by participation in the labor force, industrial branch, and employment status of husband.	113

VOLUME 18. CHILDREN AND YOUTHS IN FAMILIES. (10 PERCENT SAMPLE.)

TABLES

Number	Title	Page
1.	Unmarried children in families by age and type of family.	34
2.	Unmarried children in families, 15 years of age and	

GERMANY

	older, by age, type of participation in the labor force, industrial branch, and employment status.	35
3.	Unmarried children in families, 15 years of age and older, by participation in the labor force and employment status of the children and of the head of the family, and by type of family and number of children in the family.	38
4.	Unmarried children in families, 15 to less than 25 years of age, by age, participation in the labor force and employment status of the children and of the head of the family, and by type of family and number of children in the family.	40
5.	Unmarried children in families, 15 to less than 25 years of age, by age, participation in the labor force and employment status of the children and of the head of the family, and by type of family, number of children in the family, and expellee/refugee status.	58
6.	Selected types of families by number and age of children.	66
7.	Selected types of families by children of specified age groups.	67
8.	Married couples with children by number and age of the children, type of participation in the labor force and employment status of spouses, and industrial branch of the husband.	68
9.	Married couples with children 15 years of age and older by type of participation in the labor force of the children and total number of children in the family.	70
10.	Mothers by number and age of children, type of family, and type of participation in the labor force.	72
11.	Unmarried children in families by age, type of family, and state.	76
12.	Unmarried children in families, 15 years of age and older, by age, type of participation in the labor force, industrial branch, employment status, and state.	78
13.	Unmarried children in families, 15 years of age and older, by participation in the labor force and employment status of the children and of the head of the family, and by state.	84
14.	Unmarried children in families, 15 to less than 25 years of age, by sex, age, participation in the labor force and employment status of the children and the head of the family, and by state.	86

TABLE TITLES BY VOLUME

15. Selected types of families by number and age of the children and by state. 90
16. Selected types of families with children of specified age groups by state. 91

VOLUME 19. PRINCIPAL SOURCE OF LIVELIHOOD OF THE OLDER CITIZENS. (10 PERCENT SAMPLE.)

TABLES

Number	Title	Page
1.	Persons 45 years of age or older in single person households by marital status, principal source of livelihood, industrial branch, employment status, and sex.	38
2.	Persons 45 years of age or older in single person households by marital status, principal source of livelihood, industrial branch, employment status, sex, and expellee/refugee status.	40
3.	Persons 45 years of age or older in single person households by commune size class, principal source of livelihood, industrial branch, employment status, and sex.	44
4.	Persons 45 years of age or older in multi-person households by marital status, principal source of livelihood, industrial branch, employment status, sex, and type of household.	48
5.	Persons 45 years of age or older in multi-person households by commune size-class, principal source of livelihood, industrial branch, employment status, sex, and type of household.	52
6.	Persons 45 years of age or older in institutions by marital status, principal source of livelihood, and sex.	60

VOLUME 20. RELIGIOUS AFFILIATION OF FAMILIES. (10 PERCENT SAMPLE.)

TABLES

Number	Title	Page
1.	Married couples by number of children, year of marriage group,[1] and religious affiliation of spouses.	38

---

[1] Married persons (living together or separately) were asked

2. Married couples belonging to different religious communities by number and religious affiliation of children and parents. 58
3. Married couples belonging to different religious communities by number of children, year of marriage group of parents, and religious affiliation of children and parents. 65
4. Parents with children by number of children and religious affiliation of children and parents. 66
5. Married couples by year of marriage group, religious affiliation of spouses, and state. 68
6. Married couples with children, who belong to different religious communities, by religious affiliation of children and parents, by state. 76

---

about year of marriage group. Up to the year of marriage 1961, full calendar years of marriage were shown. Persons married as of June 6, 1961, who were married between January 1, 1961, and June 5, 1961, belonged to the year of marriage 1961.

GIBRALTAR

MAIN ENTRY:

Gibraltar.
Report on the census of Gibraltar taken on 3rd October 1961.
Gibraltar [1962?]
47 p. 33cm. (NUC63-12309)

ENUMERATION DATE: October 3, 1961.

ENUMERATION BASIS: De facto; de jure.

GLOSSARY

ADMINISTRATIVE, GEOGRAPHICAL, AND POLITICAL UNITS

Gibraltar is a Crown Colony administered by a governor with the assistance of an Executive Council and a Legislative Council.

TABLE OF DEMOGRAPHIC CONCEPTS

Appendix	United Kingdom Terminology
ECONOMIC CHARACTERISTICS	
C	Industrial classification Industry[1]
...	Occupational classification Occupations[2]

---

[1] In the table titles, "industry" designates tabulations of either of the two levels of the United Kingdom industrial classification which are employed: "Orders" or "Minimum list headings." These terms correspond to the ISIC 1958 as follows: "Orders" are "Divisions" and "Major groups"; "Minimum list headings" are "Groups."

[2] The classification by occupation in Table 12 does not conform to the ISCO 1958, but to the "Guide to Occupational Classification" used by the Ministry of Labour and National Service in the United Kingdom to register workers for employment (Great Britain, Report on the Census of Gibraltar Taken on 3rd October, 1961, p. 10).

GIBRALTAR

TABLES

Number	Title	Page
1.	The population in the years 1753 to 1961.	13
2.	Composition of total civil population.[1]	14
3.	Distribution of resident civil population by age groups and sexes [data for 1951 and 1961].	15
3A.	Distribution of resident civil population by age groups[2] and sexes.	16
3B.	Distribution of resident British subjects by age groups[2] and sexes.	17
3C.	Distribution of alien residents and visitors by age groups and sexes.	18
3D.	Families of service personnel.	19
4.	Births and birth rates during the period 1952 to 1961.	20
5.	Deaths and death rates during the period 1952 to 1961.	20
6.	Marital status according to age groups.	21
6A.	Marital status of resident British subjects according to age groups.	22
6B.	Marital status of alien residents and visitors according to age groups.	23
7.	Birthplaces.	24
7A.	Birthplaces of resident British subjects.	25
7B.	Birthplaces of alien residents and visitors.	26
8.	Nationalities.	27
9.	Religious distribution of the civil population.	28
10.	Physical and mental infirmities among the resident population.	29
11.	Housing administered by government.	30
12.	Distribution of the population by occupations.	31
13.	Distribution of the population by industries.	40
14.	Comparison with police figures.	43
15.	Emigration from Gibraltar.	43

[1] The total civil population includes permanent and temporary residents, plus transients and service families. The resident civil population includes only those persons who are permanent residents of Gibraltar.

[2] Data are provided by single years of age, 1-100.

GREAT BRITAIN

MAIN ENTRIES AND VOLUME NOTES:

ENGLAND AND WALES

Great Britain. General Register Office.
 Census 1961, England and Wales; age, marital condition and general tables. London, 1964.
  xv, 108 p. 33cm. (65-68818)

Great Britain. General Register Office.
 Census 1961, England and Wales; birthplace and nationality tables. London, 1964.
  xi, 70 p. 33cm. (65-50157)

Great Britain. General Register Office.
 Census 1961, England and Wales; Commonwealth immigrants in the conurbations. London, 1965.
  xviii, 116 p. 33cm. (66-57739)

Great Britain. General Register Office.
 Census 1961, England and Wales; county reports. London, 1963-1964.
  62 v. 33cm. (63-6098rev.)

  Anglesey. 1963. 46 p.
  Bedfordshire. 1963. 72 p.
  Berkshire. 1964. 94 p.
  Breconshire. 1963. 50 p.
  Buckinghamshire. 1964. 88 p.
  Caernarvonshire. 1963. 58 p.
  Cambridgeshire. 1964. 44 p.
  Cardiganshire. 1963. 48 p.
  Carmarthenshire. 1963. 56 p.
  Cheshire. 1964. 172 p.
  Cornwall and the Isles of
    Scilly. 1964. 85 p.
  Cumberland. 1963. 76 p.
  Denbighshire. 1963. 55 p.
  Derbyshire. 1964. 106 p.
  Devon. 1964. 143 p.
  Dorset. 1963. 78 p.
  Durham. 1963. 169 p.
  East Suffolk. 1964. 88 p.
  East Sussex. 1964. 110 p.
  Essex. 1963. 247 p.
  Flintshire. 1963. 50 p.
  Glamorgan. 1963. 124 p.
  Gloucestershire. 1964. 114 p.
  Hampshire. 1963. 141 p.
  Herefordshire. 1964. 61 p.
  Hertfordshire. 1963. 141 p.
  Huntingdonshire. 1964. 52 p.
  Isle of Ely. 1964. 48 p.
  Isle of Wight. 1963. 44 p.
  Kent. 1963. 193 p.
  Lancashire. 1964. 432 p.
  Leicestershire. 1964. 88 p.
  Lincolnshire (Parts of
    Holland). 1964. 44 p.
  Lincolnshire (Parts of
    Kesteven). 1964. 52 p.
  Lincolnshire (Parts of
    Lindsey). 1964. 108 p.
  London. 1963. 268 p.

Merionethshire. 1963. 48 p.
Middlesex. 1963. 192 p.
Monmouthshire. 1963. 96 p.
Montgomeryshire. 1963. 50 p.
Norfolk. 1964. 113 p.
Northamptonshire. 1964. 98 p.
Northumberland. 1963. 122 p.
Nottinghamshire. 1964. 104 p.
Oxfordshire. 1964. 86 p.
Pembrokeshire. 1963. 54 p.
Radnorshire. 1963. 46 p.
Rutland. 1964. 42 p.
Shropshire. 1964. 76 p.
Soke of Peterborough. 1964. 42 p.
Somerset. 1964. 115 p.
Staffordshire. 1963. 177 p.
Surrey. 1963. 189 p.
Warwickshire. 1963. 135 p.
Westmorland. 1963. 46 p.
West Suffolk. 1964. 55 p.
West Sussex. 1964. 84 p.
Wiltshire. 1964. 82 p.
Worcestershire. 1963. 108 p.
Yorkshire East Riding. 1964. 88 p.
Yorkshire North Riding. 1964. 110 p.
Yorkshire West Riding. 1963. 300 p.

Great Britain. General Register Office.
Census 1961, England and Wales; education tables. London, 1966.
xx, 111 p. 33cm. (66-78333)

Great Britain. General Register Office.
Census 1961, England and Wales; fertility tables. London, 1966.
xxii, 316 p. 33cm. (67-77818)

Great Britain. General Register Office.
Census 1961, England and Wales; Greater London tables. London, 1966.
xvi, 68 p. 33cm. (66-67787)

Great Britain. General Register Office.
Census 1961, England and Wales; household composition, national summary tables. London, 1966.
[2], xv, 2-18 p. 33cm. (68-76837)
[Excluded; preliminary data.]

Great Britain. General Register Office.
Census 1961, England and Wales; household composition tables. London, 1966.
xxxiv, 352 p. 33cm. (NUC68-45286)

Great Britain. General Register Office.
Census 1961, England and Wales; housing, national summary tables. London, 1964.
v. 33cm. (65-50158)
[Excluded.]

Great Britain. General Register Office.
Census 1961, England and Wales; housing tables. London, 1964-
    v. 33cm. (65-56351)
    [Excluded.]

Great Britain. General Register Office.
Census 1961, England and Wales; index of place names. London, 1965.
    2 v. (iv, 1324 p.) 33cm. (66-39135)
    [Excluded; no tables.]

Great Britain. General Register Office.
Census 1961, England and Wales; industry tables. London, 1966.
    2 v. in 1 (lii, 371 p.) 33cm. (66-74784)

Great Britain. General Register Office.
Census 1961, England and Wales; migration, national summary tables. London, 1965.
    2 v. (ix, 59 p.) 33cm. (66-56734)
    [Excluded; preliminary data.]

Great Britain. General Register Office.
Census 1961, England and Wales; migration tables. London, 1966.
    xxix, 358 p. 33cm. (66-68232)

Great Britain. General Register Office.
Census 1961, England and Wales; occupation and industry, national summary tables. London, 1965.
    xv, 9 p. 33cm.
    [Excluded; preliminary data.]

Great Britain. General Register Office.
Census 1961, England and Wales; occupation, industry, socio-economic groups. London, 1965-1966.
    58 v. 33cm. (66-73668)

    Anglesey. 1966. 11 p.   Cumberland. 1965. 15 p.
    Bedfordshire. 1965. 15 p.   Denbighshire. 1966. 11 p.
    Berkshire. 1965. 15 p.   Derbyshire. 1966. 15 p.
    Breconshire. 1966. 11 p.   Devon. 1966. 19 p.
    Buckinghamshire. 1965. 15 p.   Dorset. 1966. 15 p.
    Cambridgeshire. 1965. 15 p.   Durham. 1966. 29 p.
    Caernarvonshire. 1966. 11 p.   Essex. 1965. 51 p.
    Cardiganshire. 1966. 11 p.   Flintshire. 1966. 11 p.
    Carmarthenshire. 1966. 11 p.   Glamorgan. 1966. 27 p.
    Cheshire. 1966. 29 p.   Gloucestershire. 1966. 17 p.
    Cornwall & Isles of   Hampshire. 1966. 27 p.
        Scilly. 1966. 11 p.   Herefordshire. 1966. 11 p.

GREAT BRITAIN

    Hertfordshire. 1965. 29 p.
    Huntingdonshire. 1966. 11 p.
    Isle of Ely. 1966. 11 p.
    Isle of Wight. 1966. 11 p.
    Kent. 1965. 31 p.
    Lancashire. 1966. 69 p.
    Leicestershire. 1966. 15 p.
    Lincolnshire. 1966. 29 p.
    London. 1965. 73 p.
    Merionethshire. 1966. 11 p.
    Middlesex. 1965. 49 p.
    Monmouthshire. 1966. 15 p.
    Montgomeryshire. 1966. 11 p.
    Norfolk. 1966. 15 p.
    Northamptonshire. 1966. 15 p.
    Northumberland. 1966. 17 p.
    Nottinghamshire. 1966. 17 p.
    Oxfordshire. 1966. 15 p.
    Pembrokeshire. 1966. 11 p.
    Radnorshire. 1966. 11 p.
    Rutland. 1966. 11 p.
    Shropshire. 1966. 11 p.
    Soke of Peterborough. 1966. 15 p.
    Somerset. 1966. 17 p.
    Staffordshire. 1966. 31 p.
    Suffolk. 1966. 17 p.
    Surrey. 1965. 43 p.
    Sussex. 1965. 31 p.
    Warwickshire. 1966. 27 p.
    Westmorland. 1966. 11 p.
    Wiltshire. 1966. 15 p.
    Worcestershire. 1966. 17 p.
    Yorkshire East Riding. 1966. 15 p.
    Yorkshire North Riding. 1966. 17 p.
    Yorkshire West Riding. 1966. 47 p.

Great Britain. General Register Office.
    Census 1961, England and Wales; occupation tables. London, 1966.
    xiv, 267 p. 33cm.   (68-106402)

Great Britain. General Register Office.
    Census 1961, England and Wales; population, dwellings, households. London, 1962-
    v. 33cm.   (NUC64-22285)
    [Excluded; preliminary data.]

Great Britain. General Register Office.
    Census 1961, England and Wales; preliminary report. London, 1961.
    iii, 80 p. 33cm.   (66-56727)
    [Excluded; preliminary data.]

Great Britain. General Register Office.
    Census 1961, England and Wales; socio-economic group tables. London, 1966.
    xxii, 109 p. 33cm.   (NUC67-37415)

Great Britain. General Register Office.
    Census 1961, England and Wales; usual residence tables. London, 1964.
    xiii, 47 p. 33cm.   (65-68819)

MAIN ENTRIES AND VOLUME NOTES

Great Britain. General Register Office.
Why a census? London, 1961.
30 p. 21cm. (NUC64-69744)
[Excluded; no tables.]

Great Britain. General Register Office.
Census 1961, England and Wales; workplace tables. London, 1966.
xxiv, 310 p. 33cm. (66-75052)

GREAT BRITAIN

Great Britain. General Register Office.
Census 1961, Great Britain; scientific and technological qualifications. London, 1962.
xvii, 18 p. 33cm. (62-53263)

Great Britain. General Register Office.
Census 1961, Great Britain; summary tables. London, 1966.
lii, 158 p. 33cm. (NUC67-105473)

JERSEY, GUERNSEY, AND ADJACENT ISLANDS

Great Britain. General Register Office.
Census 1961. Report on Jersey, Guernsey and adjacent islands. London, 1966.
xiv, 83 p. 33cm. (66-73142)

MAN, ISLE OF

Great Britain. General Register Office.
Census 1961, Report on Isle of Man. Part I. Population and housing. London, 1965. ix, 28 p. 33cm. Part II. Migration, economic activity, and other topics. London, 1966. x, pp. 29-48. 33cm.

WALES

Great Britain. General Register Office.
Census 1961, Wales (including Monmouthshire); report on Welsh-speaking population. London, 1962.
xii, 33 p. (NUC64-22310)

ENUMERATION DATE: April 23/24, 1961.

ENUMERATION BASIS: De facto.

GREAT BRITAIN

GLOSSARY

ADMINISTRATIVE, GEOGRAPHICAL, AND POLITICAL UNITS

Level	Vernacular Term
National	Great Britain
Intermediate	...
Local	County

	Local authority areas:
	  Administrative county
	  County, metropolitan, and municipal boroughs
	  Urban and rural districts
	  New towns

Other concepts:	
	Standard region
	Rural and urban aggregates
	Conurbation Conurbation center

In Great Britain, local authority areas are locally governed areas. They include administrative counties, county boroughs, metropolitan boroughs, municipal boroughs, urban districts, and rural districts.

Population figures for counties pertain either to the administrative county (excluding any county boroughs) or to the entire county (administrative county and any associated county boroughs), as explicitly stated in each case.

Administrative counties are the geographical counties excluding the county boroughs.

County boroughs are large boroughs which have a population of at least 75,000. Boroughs can become county boroughs only by a special Act of Parliament. Each county borough has a single council responsible for all services within its area.

New towns are urban centers created by the New Towns Act in 1946 to encourage the decentralization of population and industry, which was concentrated in and around London and other large cities. The Act specified the creation of new towns at usually underdeveloped sites in England and Wales, each with its own development corporation financed by the government, a self-contained local administration, and an admixture of population.

Standard regions are regional groupings of counties.

GLOSSARY

Rural and urban aggregates are local authority areas grouped according to whether they are within conurbations, urban areas, or rural districts. Urban areas are also grouped according to the size of their resident population estimated on the census day (23rd April 1961). Urban areas include boroughs and urban districts which, together with rural districts, are defined by the Local Governments Acts.

Conurbations are vast metropolitan districts, surrounded by aggregations of urban communities.

Conurbation centers are the central areas of conurbations, containing the principal concentration of administrative and commercial offices, major shopping streets, public buildings, terminals, etc. They are characterized by relatively low residential populations, high concentrations of commuting workers, and often traffic congestion and parking problems.

TABLE OF DEMOGRAPHIC CONCEPTS

The demographic concepts which appear in the following table correspond to the recommendations of the European Conference of Statisticians, unless otherwise noted. They are defined and discussed in the appendixes and in footnotes. Terms in brackets are not used in the table titles, but they are included here in order to give logical completeness to the various classifications in the table.

Appendix	English Table Term	United Kingdom Terminology
CULTURAL, EDUCATIONAL, AND PERSONAL CHARACTERISTICS		
...	Nationality or citizenship[1]	Same
ECONOMIC CHARACTERISTICS		
A	Economic activity or classification by economic position	Same
	Economically active	Same
	Economically inactive	Same
B	Employment status	Same
C	Industrial classification[2]	Same
	Division(s)	Orders
	Major group(s)	Orders
	Group(s)	Minimum list headings
	Subgroup(s)	Sub-divisions

---
[1][All footnotes are at the end of the Table.]

GREAT BRITAIN

D	Occupational classification[3]
	   Major group(s)
	   Minor group(s)
	   [Unit groups][4]
		Same
		   Orders
		   Unit groups
		   ...

...	...	Social class[5]
E	Socio-economic category	Socio-economic group[6]
...	...	Status group[7]

HOUSEHOLD AND FAMILY CHARACTERISTICS

F	Household
	   Private household[8]
	      Households sharing
	         dwellings[9]
	      Types of households[10]
	      ...
	      Household supporter[12]
	
	      Head of household[13]
	   Institutional household[14]
	   Family[15]
	      Head of the family[16]
	      Dependent children[17]
		Same
		   Same
		      Same
		
		      Same
		      Tenure of dwellings[11]
		      Chief economic sup-
		        porter
		      Same
		   Non-private household
		   Same
		      Same
		      Same	

POPULATION BASE FOR THE CENSUS[18]

...	...	Estimated resident popula- tion (census definition)[19]
...	...	Estimated resident popula- tion (estimate defini- tion)[20]

---

[1] The classification by nationality or citizenship applies only to those persons not born in Great Britain or Northern Ireland. The census questionnaire posed three questions: (1) of Commonwealth citizens, their country of citizenship--e.g., United Kingdom, India, Canada; (2) of citizens of the U.K. and the Colonies, whether citizenship was by birth, descent, naturalization, registration, or marriage; (3) of other persons, their nationality--e.g., Italian, Polish.

[2] In the table titles, "industry" designates tabulations of any and all levels of industrial classification. The U.K. system of industrial classification corresponds to the ISIC 1958 as follows: "Orders" are "Divisions" and "Major groups"; "Minimum list headings" are "Groups"; and "Sub-divisions" are "Sub-groups."

[3] In the table titles, "occupation" designates tabulations of either the ISCO 1958 major and minor groups together, or of the major groups alone. In the U.K. system of occupational

classification, "Orders" are "Major groups," and "Unit groups" are "Minor groups." For a detailed discussion of the comparability of the U.K. occupational classification system to that of the ISCO 1958, see Great Britain, Classification of Occupations, 1960, p. vi.

[4] U.K. use of the term "Unit group" should not be confused with the ISCO 1958 definition of the term.

[5] The classification by social class groups the large number of unit groups of the occupational classification into several broad categories, as follows: (1) professional occupations, (2) intermediate occupations, (3) skilled occupations, (4) partly skilled occupations, (5) unskilled occupations.

[6] The classification by socio-economic group is similar, but not identical, to that recommended by the European Conference of Statisticians, described in Appendix E. Full definitions of the socio-economic groups in terms of occupation and employment status are given in Great Britain, Classification of Occupations, 1960.

[7] In the table titles, "status group" designates the economically active and inactive populations in terms of employment status and economic position, as follows:

Economically active
  A. Out of employment
      a. Sick
      b. Others
  B. In employment
      a. Self-employed
          1. Without employees
          2. With employees
               a. Large establishments
               b. Small establishments
      b. Managers
          1. Large establishments
          2. Small establishments
      c. Foremen and supervisors
      d. Apprentices, articled clerks, and formal trainees
      e. Professional employees
      f. Other employees
      g. Family workers
      h. Part-time workers

Economically inactive
  A. Institution inmates
  B. Retired persons
  C. Students
  D. Other persons economically inactive: all persons over the age of 15 without paid occupation or former occupation, including housewives.

8. Private households comprise one person living alone or a group of persons living together, taking meals together, and benefiting from common housekeeping. A person or persons living, but not boarding, in a house, flat, etc., are treated as a separate household; but a person who is usually provided with at least one meal a day by the household in which he lives is considered part of that household. A household must have exclusive use of at least one room. If two people share one room and do not have exclusive use of at least one other room, they are treated as one household.

9. Households sharing dwellings are households which do not occupy a whole dwelling. They are divided into two groups:

"E": Those which have exclusive use of both a kitchen stove or range and a kitchen sink; and
"N": Those which do not have exclusive use of both facilities.

10. Types of households are households classified by family or non-family group, as follows:

a. No family, one person (and any domestic servants).
b. One family comprising a married couple and no others.
c. One family (married couple or lone parent) with children but with no others.
d. Other family households.
e. Two or more families with or without others.
f. No family, two or more related persons, with or without others.
g. No family, two or more related persons, or domestic servants only.

11. Tenure of dwellings refers to the nature of the dwelling's occupancy. When only one household is present, the dwelling is assigned to the tenure of that household. When more than one household is present and the tenure categories of the households differ, the dwelling is assigned to the tenure of the household with the highest ranking tenure category in the dwelling. Tenure categories are ranked in this order:

a. Owner-occupied.
b. Rented with a farm, shop, or other business premises.
c. Held by virtue of employment.
d. Rented from a Local Authority or New Town Corporation.
e. Rented from a private person or company; unfurnished.
f. Rented from a private person or company; furnished.

12. The chief economic supporter ("chief") is selected from among those members of the household who are not boarders, or employees of the head of the household, or unrelated to the head of the household, or under the age of 15. The chief is selected by applying the following set of rules:

GLOSSARY

a. Employment status is considered first. Those employed full time (or whose hours worked per week are not stated), or out of employment, are selected before those employed part time, who are selected before those retired, who are selected before any others.
b. Among those selected above, position in the family is considered next, family heads being selected before any other member of the family, or before persons not in the family.
c. Among those selected by rules (a) and (b), sex is considered next, males being selected before females.
d. Among those selected by rules (a), (b), and (c), age is considered next, older persons being selected before younger.

If two or more people are finally selected according to these rules, the person whose name appears first on the census schedule is selected as chief.

[13] Head of household is the person so designated on the census schedule by the householder making the return.

[14] Non-private households include people in hotels, boarding houses, and institutions, or people living in establishments with some functional purpose other than that of providing food or otherwise satisfying domestic convenience. They include people in defense establishments, on ships at anchor or on coastwise voyages, and a small miscellaneous element.

[15] A family is a married couple living alone or with their never-married child or children (of any age). A family may also be a lone parent with a never-married child or children. A lone parent is a married parent whose spouse does not reside in the same household, or any single, widowed, or divorced parent.

[16] The head of the family is the husband in the case of a married couple, and otherwise the lone parent.

[17] Dependent children are children under 15 years of age, or persons of any age pursuing full-time education. A child is a never-married person of any age (including stepchildren or adopted children) living in the same household with at least one parent, or at least one grandparent when there are no parents, or at least one great-grandparent when there are no parents or grandparents.

[18] The census is based on the de facto (or enumerated) population. However, occasionally the tables refer to an estimated resident population, or to an estimated resident mid-year population (see footnotes 19 and 20 below).

[19] The estimated resident population (census definition) is made up of three elements:
  a. The civilian resident population, based on the statement of usual residence on the census schedule. This includes, in the area where they are enumerated, people who say they usually reside outside England and Wales.
  b. Members of the Armed Forces enumerated at stations or resident in married quarters within the area.
  c. Members of the Armed Forces who have homes within the area and who are in England and Wales outside stations. When substantial numbers of non-civilians absent on leave from stations in an area gave (incorrectly) the station instead of their home address as their usual residence, the figures were adjusted to correct that error.

This definition is intended to improve the response on the census schedule and to give the most useful figures for migration tables. It differs, however, from the definition of resident population that is used for the Registrar General's annual population estimates.

[20] The estimated resident population (estimate definition) is obtained by adjusting the estimated resident population (census definition) as follows:
  a. Deducting members of the Armed Forces enumerated at stations, and those resident in married quarters, in the area;
  b. Adding the estimated number of members of the Armed Forces stationed in the area at census date (those in married quarters being treated as stationed there);
  c. Adding students and children usually resident in colleges, schools, and university lodgings in the area;
  d. Excluding from their home areas students and children included in (c) and also those members of the Armed Forces enumerated in the area or away from their stations or married quarters. The numbers to be removed from each area are not known and are estimated arbitrarily. The assumption of an even proportional spread is unlikely to produce serious errors. [When the school instead of the home address was given (incorrectly) as the usual residence of substantial numbers of children enumerated at school, the figures were usually corrected by reducing (c) and not by adjusting the estimated resident population (census definition).]

This estimated resident population (estimate definition) is comparable with the annual mid-year estimates.

TABLE TITLES BY VOLUME

CENSUS 1961. ENGLAND AND WALES; AGE, MARITAL CONDITION AND GENERAL TABLES.[1]

TABLES

Number	Title	Page
	Population	
1.	Population by sex and intercensal changes, 1801-1961.	1
2.	Estimated population by sex as at the middle of each year, 1801-1961.	2
3.	Populations at selected censuses, 1801-1961 and 1961 acreage.[2]	3
4.	Proportional distribution of population by area at selected censuses, 1801-1961 and 1961 acreage.[2]	5
5.	Average annual rates of change of population between selected censuses, 1801-1961.[2]	7
6.	Number of areas by type and size of population.	9
7.	Census populations, 1951-1961: intercensal changes: density in 1961.[3]	10
	Age and Marital Condition	
8.	Age (single years) and marital condition.	27
9.	Age: numbers at successive censuses, 1841-1961 (thousands).	30
10.	Age: proportions at successive censuses, 1841-1961, per 10,000 persons.	31
11.	Age: proportions at successive censuses, 1841-1961, per 100,000 of each sex.	32
12.	Age and marital condition at successive censuses, 1851-1961 (thousands).	33

---

[1] No sampling is involved. Unless otherwise noted, the area for which statistics are given is England and Wales.

[2] England and Wales, regions, conurbations, administrative aggregates, counties.

[3] England and Wales, regions, conurbations, urban/rural aggregates, counties, local authority areas.

GREAT BRITAIN

13.	Marital condition proportions at successive censuses, 1851-1961 (per 1,000 in each sex-age group).	35
14.	Marital condition proportions and sex ratios by age, 1921-1961.	37
15.	Age.[4]	38
16.	Age and marital condition.[4]	43
17.	Age proportions per 1,000 of each sex, 1951 and 1961.[4]	51
18.	Marital condition proportions (per 1,000 in each sex-age group).[4]	55
19.	Sex, age and marital condition indices.[5]	57

Non-private Households

20.	Population enumerated in private and non-private households.[6]	90
21.	Institutions.[6]	94
22.	Hotels and boarding houses: number, rooms and population.[6]	102
23.	Population enumerated in and outside private households by sex and age.	104
24.	Institutions: age and marital condition of inmates.	105
25.	Children's homes: age of inmates.	107
26.	Hotels and boarding houses: age and marital condition of guests.	108

---

[4] England and Wales, regions, conurbations, urban/rural aggregates.

[5] England and Wales, regions, conurbations, urban/rural aggregates, counties, local authority areas, conurbation centers, new towns.

[6] England and Wales, regions, conurbations, urban/rural aggregates, counties.

CENSUS 1961. ENGLAND AND WALES; BIRTHPLACE AND NATIONALITY TABLES.[1]

TABLES

Number	Title	Page
1.	Birthplaces and nationalities of the whole population.[2]	1
2.	Birthplaces, nationalities and citizenships of residents of England and Wales born outside the British Isles.[3]	4
3.	Nationalities and citizenships of residents of England and Wales born outside the British Isles or with birthplace not stated.[3]	39
4.	Birthplaces: proportions per 10,000 of each sex.[4]	55
5.	Commonwealth citizens and citizens of the Irish Republic resident in, but born outside, England and Wales: age and marital condition by country of birth.	65
6.	Aliens resident in England and Wales: age and marital condition by country of nationality.	68
7.	Birthplaces of the population at selected censuses, 1851-1961.	70
8.	Birthplaces of the population at selected censuses, 1851-1961 (proportions per 100,000 persons).	70
9.	Natives of England and Wales enumerated at selected censuses, 1851-1961.[5]	70

---

[1] No sampling is involved. Unless otherwise noted, the area for which statistics are given is England and Wales.

[2] England and Wales, regions, conurbations.

[3] England and Wales, regions, conurbations, counties with 2,000 or more born outside the British Isles.

[4] England and Wales, regions, conurbations, counties, county boroughs, metropolitan boroughs, urban areas with populations of 50,000 or more, county remainders, new towns.

[5] Countries of the British Isles.

GREAT BRITAIN

CENSUS 1961. ENGLAND AND WALES; COMMONWEALTH IMMIGRANTS IN THE CONURBATIONS.[1]

TABLES

Number	Title	Page
	Persons Born in Specified Commonwealth Countries and Colonial Territories	
A.1.	Sex by age, by marital condition.	1
A.2.	Sex by age, by marital condition, by birthplace.[2]	2
A.3.	Industry by sex, by age, by birthplace.[3]	4
A.4.	Occupation by sex, by age, by birthplace.[3]	40
A.5.	Socio-economic group of persons in employment by sex, by age, by birthplace.[3]	86
A.6.	Industry[4] by occupation, by sex.	92
A.7.	Industrial status by sex, by birthplace.	97
A.8.	Out of employment. Occupation by sex by reason for being out of employment.[3]	98
A.9.	Persons qualified in science and technology by age, socio-economic group, and industry.	101
A.10.	Duration of residence by sex, by marital condition, by socio-economic group.	104
A.11.	Population aged 15 and over by terminal education age, by sex.	105
A.12.	Persons enumerated in non-private households by sex, by age, by type of institution, by birthplace.[2]	106

---

[1] Ten percent sample. The term "Commonwealth immigrants" is used for persons who gave as their birthplace Jamaica, other Commonwealth territories in the Caribbean, India, Pakistan, Commonwealth countries in Africa (excluding the Union of South Africa), Cyprus, or Malta. The tables include only those Commonwealth immigrants who were resident in the six conurbations in England and Wales. Unless otherwise noted, the area for which statistics are given is the six conurbations combined.

[2] Six conurbations combined and Greater London conurbation.

[3] Six conurbations combined, Greater London conurbation, and remainder.

[4] In the tables on industry in this volume, unlike the general tables on industry for the whole population, people are allocated to the area in which they live instead of the area in which they work.

TABLE TITLES BY VOLUME

Households with Head or Wife of Head Born in Specified Commonwealth Countries and Colonial Territories

B.1. Private households by size, by rooms occupied, by sharing of dwellings, by household type. 108

B.2. Private households by size, by household type, by birthplace of head or spouse.[2] 110

B.3. Private households by tenure, by density of occupation (persons per room), by birthplace of head or spouse.[5] 112

B.4. Households by number of children and children by age, by birthplace of head or spouse.[2] 114

B.5. Households with children by type of marriage, by birthplace of head or spouse and children, by type of marriage, by (child's) birthplace, by descent.[2] 115

CENSUS 1961. ENGLAND AND WALES. COUNTY REPORTS.
[62 reports; see listing on pp. 125-126 above.]

Structure of the County Reports

The table numbers and table titles for all the County Reports are identical. Page locations, however, vary from volume to volume. Hence, it is necessary to consult the Tables list and the Page Location Matrix which follow in order to locate desired data.

TABLES[6]

Number   Title

Population

1. Population 1801-1961 and intercensal variations.[7]
2. Population 1931-1961 and intercensal variations.[8]
3. Acreage, population, private households and dwellings.[9]

---

[5] Six conurbations separately.

[6] For all counties, statistics for the administrative county with any county boroughs are included in every table except Tables 5 and 17.

[7] County.

[8] Local authority areas.

[9] Local authority areas, wards, civil parishes in rural districts,

GREAT BRITAIN

Boundary Changes and Detached Parts

4. Intercensal changes of boundary (between 8th April, 1951 and 23rd April, 1961).[10]
5. Detached parts of local authority areas and civil parishes: acreage and population.[10]

Sex, Age, and Marital Condition

6. Age and marital condition.[11]
7. Age--single years under 21.[11]

Birthplace and Nationality

8. Birthplace and nationalities of the whole population.[12]
9. Residents born outside England and Wales, by nationality and citizenship, and visitors from outside England and Wales.[13]
10. Birthplaces, nationalities, and citizenships of residents of England and Wales born outside the British Isles.[14]

Housing and Households

11. Dwellings by building type, rooms and household spaces.[11]
12. Buildings and dwellings.[15]
13. Private households by size, rooms occupied, and sharing of dwellings.[11]

---

conurbation centers, new towns. London: Local authority areas, wards, conurbation center.

[10] Local authority areas, civil parishes.

[11] Local authority areas, conurbation centers, new towns. London: Local authority areas, conurbation center.

[12] County boroughs, urban areas with populations of 50,000 or more, county remainder, new towns. London: Local authority areas.

[13] Local authority areas, new towns. London: Local authority areas.

[14] Urban areas with 2,000 or more born outside the British Isles. London: Local authority areas with 2,000 or more born outside the British Isles.

[15] County boroughs, urban areas with populations of 50,000 or more, conurbation centers, new towns. London: Local authority areas, conurbation center.

14. Private households by size, rooms occupied, and type of building.[16]
15. Dwellings by tenure and rooms.[16]
16. Private households by size, rooms occupied, sharing of dwellings, and tenure.[16]
17. Private households, rooms and persons, by tenure.[17]
18. Private households by density of occupation (persons per room).[11]
19. Population in all private households by density of occupation (persons per room).[11]
20. Private households by type of building and tenure (proportions per 1,000).[16]
21. Households sharing a dwelling as a percentage of all households in each tenure category.[16]

Household Arrangements

22. Dwellings by availability of certain household arrangements.[11]
23. Private households by availability of certain household arrangements.[11]

Non-private Households

24. Population enumerated in private and non-private households.[16]
25. Institutions.[11]
26. Hotels and boarding houses. Number, rooms, and population.[18]

Old People

27. One- and two-person households containing persons of pensionable age by sex/condition combination.[11]

---

[16] County boroughs, urban areas with populations of 50,000 or more, county aggregates, conurbation centers, new towns. London: Local authority areas, conurbation center.

[17] Urban areas with populations of less than 50,000, rural districts.

[18] County boroughs, areas with 500 or more hotel rooms, county remainder. London: County, local authority areas with 500 or more hotel rooms, county remainder.

GREAT BRITAIN

PAGE LOCATION MATRIX

Table Number	Anglesey	Bedfordshire	Berkshire	Breconshire	Buckinghamshire	Caernarvonshire	Cambridgeshire	Cardiganshire	Carmarthenshire	Cheshire	Cornwall and the Isles of Scilly
1	1	1	1	1	1	1	1	1	1	1	1
2	1	1	2	1	1	1	1	1	1	1	2
3	2	2	3	2	2	2	2	2	2	2	3
4	3	4	6	3	6	4	4	3	4	10	7
5	3	5	6	3	10	4	5	3	4	11	7
6	4	6	7	4	11	5	6	4	5	12	8
7	7	10	13	7	16	9	8	7	9	23	15
8	8	11	15	8	18	10	9	8	10	26	17
9	8	11	16	8	19	11	9	8	10	28	18
10	9	12	17	9	20	11	10	9	11	29	19
11	10	14	20	10	22	12	12	10	12	32	20
12	17	26	36	18	35	22	17	18	21	64	35
13	18	27	37	19	36	23	18	19	22	65	36
14	23	34	48	25	47	31	21	25	29	88	49
15	26	39	54	28	52	34	24	28	32	99	52
16	30	45	62	32	58	38	28	32	36	112	56
17	36	55	74	38	68	44	34	38	42	134	62
18	36	56	76	38	70	46	34	38	44	136	64
19	38	58	78	40	72	47	36	40	46	140	66
20	38	59	79	41	73	48	37	40	47	142	67
21	38	59	79	41	73	48	37	40	47	69	68
22	39	60	80	42	74	49	38	41	48	70	69
23	41	63	84	45	78	52	39	43	51	74	75
24	44	68	90	48	84	56	42	46	54	80	80
25	44	70	92	48	86	56	42	46	54	82	82
26	45	71	93	49	87	57	43	47	55	83	84
27	46	72	94	50	88	58	44	48	56	84	85

TABLE TITLES BY VOLUME

Table Number	Cumberland	Denbighshire	Derbyshire	Devon	Dorset	Durham	East Suffolk	East Sussex	Essex	Flintshire	Glamorgan
1	1	1	1	1	1	1	1	1	1	1	1
2	1	1	2	2	1	1	1	1	1	1	1
3	2	2	3	3	2	2	2	2	2	2	2
4	5	4	8	10	6	8	7	5	9	3	6
5	5	4	8	10	7	11	8	7	13	3	7
6	6	5	9	11	8	12	9	8	14	4	8
7	10	8	16	22	13	22	14	14	27	7	15
8	12	10	18	25	15	25	16	16	31	8	17
9	12	10	19	26	15	27	17	17	34	8	18
10	13	11	20	27	16	28	18	18	35	9	19
11	14	12	22	30	18	30	20	22	42	10	22
12	27	21	41	58	31	61	35	41	83	19	44
13	28	22	42	59	32	62	36	42	84	20	45
14	37	29	57	81	42	84	47	54	110	26	60
15	42	32	63	88	46	96	52	62	132	29	69
16	48	36	70	98	51	110	58	72	156	33	80
17	58	42	82	112	59	134	68	88	200	39	98
18	60	42	84	114	60	136	70	90	202	40	100
19	62	44	86	116	62	140	72	92	206	40	102
20	63	45	87	118	63	142	73	93	207	41	103
21	63	45	87	118	63	142	73	93	208	42	103
22	64	46	88	119	64	143	74	94	209	42	104
23	67	49	94	128	68	151	78	98	218	45	110
24	72	52	102	138	74	164	84	106	240	48	120
25	74	54	104	140	76	166	86	108	242	48	122
26	75	54	105	142	77	168	87	109	245	49	123
27	76	55	106	143	78	169	88	110	246	50	124

GREAT BRITAIN

Table Number	Gloucestershire	Hampshire	Herefordshire	Hertfordshire	Huntingdonshire	Isle of Ely	Isle of Wight	Kent	Lancashire	Leicestershire
1	1	1	1	1	1	1	1	1	1	1
2	1	1	1	2	1	1	1	1	2	1
3	2	2	2	3	2	2	2	2	4	2
4	7	7	6	7	4	3	3	9	18	6
5	10	10	6	11	5	3	3	12	22	7
6	11	11	7	12	6	4	4	13	23	8
7	18	18	10	21	9	7	6	26	51	13
8	20	21	12	24	10	8	7	30	59	15
9	21	22	13	25	11	9	8	32	62	16
10	22	23	13	26	11	9	9	33	64	17
11	26	28	14	28	12	10	10	38	72	20
12	45	52	23	51	20	18	16	73	158	35
13	46	53	24	52	21	19	17	74	160	36
14	60	70	32	70	27	25	21	100	222	47
15	67	80	35	80	30	28	24	113	252	52
16	76	92	39	92	34	32	28	128	284	58
17	90	112	45	112	40	38	34	154	344	68
18	92	114	46	114	40	38	34	156	348	70
19	94	116	48	116	42	40	36	158	356	72
20	95	117	49	117	43	40	36	159	360	73
21	96	118	50	118	44	40	37	159	361	73
22	97	119	51	119	45	41	37	160	362	74
23	102	125	54	126	47	43	39	171	384	78
24	110	136	58	136	50	46	42	186	420	84
25	112	138	60	138	50	46	42	188	424	86
26	113	140	60	140	51	47	43	191	429	87
27	112	141	61	141	52	48	44	192	430	88

TABLE TITLES BY VOLUME

Table Number	Lincolnshire (Parts of Holland)	Lincolnshire (Parts of Kesteven)	Lincolnshire (Parts of Lindsey)	London	Merionethshire	Middlesex	Monmouthshire	Montgomeryshire	Norfolk	Northamptonshire
1	1	1	1	1	1	1	1	1	1	1
2	1	1	1	1	1	1	1	1	1	1
3	2	2	2	2	2	2	2	2	2	2
4	3	5	8	8	3	5	5	3	10	7
5	3	5	9	8	3	5	5	3	14	9
6	4	6	10	9	4	6	6	4	15	10
7	6	9	16	20	7	15	12	7	22	16
8	7	10	18	22	8	17	14	8	24	18
9	8	11	19	26	8	19	15	8	25	19
10	9	12	20	27	9	20	16	9	26	20
11	10	13	22	46	10	32	18	10	28	22
12	16	21	40	78	18	58	35	18	47	39
13	17	22	41	80	19	59	36	19	48	40
14	21	27	54	96	25	73	49	25	63	52
15	24	30	61	127	28	94	55	28	69	58
16	28	34	70	160	32	116	62	32	77	65
17	34	40	84	---	38	158	74	38	89	77
18	34	40	86	222	38	158	76	38	90	78
19	36	42	88	224	40	160	78	40	92	80
20	36	43	89	225	40	161	79	41	93	81
21	37	43	90	226	40	161	79	41	93	81
22	37	44	91	227	41	162	80	42	94	82
23	39	46	96	233	43	167	85	45	100	87
24	42	50	104	264	46	188	92	48	108	94
25	42	50	106	266	46	190	94	48	110	96
26	43	51	107	268	47	191	95	49	112	97
27	44	52	108	269	48	192	96	50	113	98

GREAT BRITAIN

Table Number	Northumberland	Nottinghamshire	Oxfordshire	Pembrokeshire	Radnorshire	Rutland	Shropshire	Soke of Peterborough	Somerset	Staffordshire
1	1	1	1	1	1	1	1	1	1	1
2	1	2	1	1	1	1	1	1	1	1
3	2	3	2	2	2	2	2	2	2	2
4	6	7	6	5	3	3	6	3	8	8
5	15	8	10	5	3	3	6	3	11	9
6	16	9	11	6	4	4	7	4	12	10
7	23	15	15	9	7	6	13	6	20	20
8	25	17	17	10	8	7	15	7	23	23
9	26	18	18	10	8	8	16	8	24	25
10	27	19	19	11	9	9	17	9	25	26
11	30	22	22	12	10	10	18	10	28	30
12	51	39	35	21	17	15	31	15	49	63
13	52	40	36	22	18	16	32	16	50	64
14	67	52	45	28	24	19	43	19	67	86
15	74	59	50	31	27	22	46	22	73	99
16	82	68	57	35	30	26	50	26	80	114
17	96	82	67	41	36	32	56	32	90	140
18	98	84	68	42	36	32	58	32	92	142
19	100	86	70	44	38	34	60	34	94	146
20	101	87	71	45	38	35	61	35	95	148
21	102	87	72	45	38	35	61	35	95	148
22	103	88	73	46	39	36	62	36	96	149
23	109	93	76	49	41	37	67	37	103	157
24	118	100	82	52	44	40	72	40	110	172
25	120	102	84	52	44	40	74	40	112	174
26	121	103	85	53	45	41	75	41	114	176
27	122	104	86	54	46	42	76	42	115	177

TABLE TITLES BY VOLUME

Table Number	Surrey	Warwickshire	Westmorland	West Suffolk	West Sussex	Wiltshire	Worcestershire	Yorkshire East Riding	Yorkshire North Riding	Yorkshire West Riding
1	1	1	1	1	1	1	1	1	1	1
2	1	1	1	1	1	1	2	2	1	2
3	2	2	2	2	2	2	3	3	2	4
4	7	7	4	5	5	6	7	7	10	17
5	8	10	4	6	6	7	9	8	10	19
6	9	11	5	7	7	8	10	9	11	20
7	19	18	7	10	12	14	16	14	19	41
8	22	20	8	11	14	16	18	16	22	47
9	24	21	9	12	15	17	19	17	23	49
10	25	22	9	13	16	18	20	18	24	51
11	34	26	10	14	18	20	22	20	26	58
12	64	48	17	22	30	34	40	34	46	118
13	65	49	18	23	31	35	41	35	47	119
14	84	63	23	29	40	46	54	46	63	164
15	101	74	26	32	46	50	61	51	68	181
16	120	87	30	36	53	55	70	58	74	200
17	154	109	36	42	65	63	84	68	84	234
18	156	110	36	42	66	64	86	70	86	238
19	158	112	38	44	68	66	88	72	88	244
20	159	113	38	45	69	67	89	73	89	247
21	159	113	39	45	69	67	90	73	90	248
22	160	114	40	46	70	68	91	74	91	249
23	167	119	41	49	74	73	96	78	98	267
24	184	130	44	52	80	78	104	84	106	290
25	186	132	44	54	82	80	106	86	108	294
26	188	134	45	54	83	81	107	87	109	297
27	189	135	46	55	84	82	108	88	110	298

GREAT BRITAIN

CENSUS 1961. ENGLAND AND WALES; EDUCATION TABLES.[1]

TABLES

Number	Title	Page
1.	Population aged 15 and over in 9 age sections classified by 16 terminal education age groups: (a) numbers and (b) proportions per 1,000.	1
2.	Population aged 15 and over in 9 age sections classified by 7 terminal education age groups.[2]	3
3.	Population aged 15 and over by socio-economic group and economically active population by occupation, each classified by age and 6 terminal education age groups.	17
4.	Population aged 25 and over by 7 terminal education age groups.[3]	67

CENSUS 1961. ENGLAND AND WALES; FERTILITY TABLES.

TABLES

Number	Title	Page
1(i).	Married women by age at census and size of family.	1
1(ii).	Widowed and divorced women by age at census and size of family.	2
2(i).	All married women: size of family by duration of current marriage and age at current marriage.	3
2(ii).	Current fertility of married women: age at current marriage by duration of current marriage and size of family.	6
3.	Women with uninterrupted first marriage: age at census by duration of marriage: number of women, number of children, mean family size, number of women infertile, and proportion infertile.	8
4.	Women with uninterrupted first marriage: size of family by age at census.	18

---

[1] Ten percent sample. The tables in this volume classify information according to the age at which full-time education at school, college, or university ceased. Unless otherwise noted, the area for which statistics are given is England and Wales.

[2] Regions, conurbations, urban/rural aggregates.

[3] Counties, local authority areas, conurbation centers, new towns.

TABLE TITLES BY VOLUME

5(i).	Women with uninterrupted first marriage: size of family by duration of marriage and age at census.	19
5(ii).	Current fertility of women with uninterrupted first marriage: by age at census, duration of marriage, and size of family.	24
6(i).	Women with uninterrupted first marriage: size of family by duration of marriage and age at marriage.	25
6(ii).	Current fertility of women with uninterrupted first marriage: age at marriage by duration of marriage and size of family.	28
7.	Women with uninterrupted first marriage: size of family by year of marriage and age at marriage.	29
8(i).	Remarried women (excluding women with uninterrupted first marriage): size of family by age at current marriage and duration of current marriage.	36
8(i). Supplement	All remarried women with current marriage, age 45 and over: size of family by age at current marriage, and duration of current marriage.	39
8(ii).	Current fertility of remarried women (excluding women with uninterrupted first marriage): age at current marriage, duration of current marriage, and size of family.	40
9(i).	Remarried women (excluding women with uninterrupted first marriage): size of family by age at first marriage and time since first marriage.	42
9(ii).	Current fertility of remarried women (excluding women with uninterrupted first marriage) by age at first marriage, time since first marriage, and size of family.	45
10(i).	Women married once only and enumerated with their husbands: size of family by age at marriage and duration of marriage.	47
10(ii).	Current fertility of women married once only and enumerated with their husbands: by age at marriage, duration of marriage, and size of family.	49
11.	Women married once only and enumerated with their husbands: size of family by husband's age at census.	50
12(i).	Ratios of the fertility of women with uninterrupted first marriage to that of all married women.	51
12(ii).	Ratios of the fertility of women married once only and enumerated with their husbands to that of women with uninterrupted first marriage.	52

12(iii).	Ratios of the fertility of remarried women (excluding women with uninterrupted first marriage) classified by current marriage to that of women with uninterrupted first marriage.	53
12(iv).	Ratios of the fertility of remarried women (excluding women with uninterrupted first marriage), classified by first marriage, to that of women with uninterrupted first marriage.	54
13.	Women with uninterrupted first marriage: fertility by age at marriage, and duration of marriage.	56
14.	Women married once only and enumerated with their husbands: fertility by age at marriage, duration of marriage, and socio-economic group of husband.	112
15.	Women married once only and enumerated with their husbands: fertility by age at marriage, duration of marriage, and difference between ages of husband and wife.	150
16.	Women married once only and enumerated with their husbands: fertility by age at marriage, duration of marriage, and industry group of husband.	164
17.	Women married once only and enumerated with their husbands: fertility by age at marriage, duration of marriage, and occupation order of husband.	190
18.	Women married once only and enumerated with their husbands: fertility by age at marriage, duration of marriage, and terminal education age of husband and wife.	244
19.	Women married once only and economically active: fertility by age at marriage, duration of marriage, and occupation group.	276
20.	Women with uninterrupted first marriage: fertility by age at marriage, duration of marriage, and citizenship and country of birth.	288
Appendix A.	Women married once only: age at census by year of marriage.	310

CENSUS 1961. ENGLAND AND WALES; GREATER LONDON TABLES.[1]

TABLES

Number	Title	Page
	Population	
2.	Population 1931-1961 and intercensal variations.	1

[1] The Greater London Council area (established in 1963) comprises

TABLE TITLES BY VOLUME

3.	Acreage and population.	2
	Sex, Age, and Marital Condition	
6.	Age and marital condition.	7
7.	Age--single years under 21.	19
	Birthplace and Nationality	
9.	Residents born outside England and Wales by nationality and citizenship, and visitors from outside England and Wales.	21
	Housing and Households	
11.	Dwellings by room and household spaces.	22
13.	Private households by size, rooms occupied, and sharing of dwellings.	34
17.	Private households, persons, and rooms by tenure.	46
18.	Private households by density of occupation (persons per room).	48
19.	Population in all private households by density of occupation (persons per room).	50
20.	Private households by type of building and tenure (proportions per 1,000).	51
	Household Arrangements	
22.	Dwellings by availability of certain household arrangements.	52
23.	Private households by availability of certain household arrangements.	58
	Non-private Households	
24.	Population enumerated in private and non-private households.	62
25.	Institutions.	64

---

the City of London and 32 London boroughs. This new area covers the former counties of London and Middlesex and parts of Essex, Hertfordshire, Kent, and Surrey. The present volume contains figures for Greater London and each of the London boroughs selected from tables in the County Reports. The table numbers used in the original series of County Reports have been retained. (Tables 1, 4, 5, etc. which appear in the County Reports but which do not apply to the Greater London area have been omitted.) Unless otherwise noted, the areas for which statistics are given are Greater London, the City of London, and London boroughs.

GREAT BRITAIN

26. Hotels and boarding houses. Number, rooms, and population.[2]   67

Old People

27. One- and two-person households containing persons of pensionable age by sex/condition combination.   68

CENSUS 1961. ENGLAND AND WALES; HOUSEHOLD COMPOSITION TABLES.

TABLES

Number	Title	Page
1.	Dwellings: rooms by household spaces (de jure) and households (de facto).[3]	1
2.	Households: persons (de jure and de facto) by rooms, by sharing of dwellings.[4]	2
3.	Households recorded as entirely absent from usual residence: persons usually resident by rooms.[3]	6
4.	Households: socio-economic group of chief economic supporter by tenure, by persons, by rooms.[4]	7
5.	Households: household type by persons.[4]	28
6.	Households: household type by persons, by children.[4]	29
7.	Households: household type by persons, by rooms.[4]	34
8.	Households: socio-economic group of chief economic supporter by household type, by persons.[4]	44
9.	Households and families by type.[5]	65
10.	Households: family type of chief economic supporter by his economic activity and household type.[4]	100
11.	Households containing persons of pensionable age: household type. Persons of pensionable age in households: household type by sex, age, marital condition, and economic activity.[4]	102
12.	One- and two-person households containing persons of	

---

[2] Areas with 500 or more hotel rooms, Greater London remainder.

[3] England and Wales (1 percent sample).

[4] England and Wales (10 percent sample).

[5] England and Wales, regions, conurbations, urban/rural aggregates, counties, local authority areas, conurbation centers, new towns (10 percent sample).

	pensionable age: household type by rooms, by sharing of dwelling.[6]	103
13.	One- and two-person households containing persons of pensionable age: sex/condition combination.[7]	132
14.	Husbands' and wives' ages in combination, by socio-economic group of husband.[4]	133
15.	Households: persons by earners, by children.[4]	140
16.	Households: persons by earners and children.[6]	142
17.	Households: socio-economic group of chief economic supporter by persons, by earners and children.[4]	148
18.	Households: socio-economic groups of chief economic supporter by persons, earners and children.[6]	152
19.	Households: industry of chief economic supporter by persons and earners.[4]	172
20.	Households: household type, by earners, by persons.[4]	173
21.	Households: household type, by persons and earners.[6]	176
22.	Households: persons by household type, earners and children.[8]	205
23.	Households containing visitors: persons in household by visitors.[4] Visitors: persons in household by type of visitor. Visitor families: persons in household by family type.	252
24.	Households containing domestic servants: domestic servants, by persons in household.[4] Domestic servants: sex and marital condition by persons in household.	253
25.	Households containing domestic servants and domestic servants' families: household type and family type.[4]	254
26.	Households containing children, persons of pensionable age, and domestic servants: household type.[6]	255

---

[6] England and Wales, regions, conurbations, urban/rural aggregates (10 percent sample).

[7] England and Wales, regions, conurbations, urban/rural aggregates (100 percent).

[8] England and Wales, counties, county boroughs, metropolitan boroughs, urban areas with populations of 50,000 or more, county remainders, new towns (10 percent sample).

GREAT BRITAIN

27.	Persons in households: household type by whether in families and economic activity.[6]	270
28.	Households with chief economic supporter born outside the British Isles: household type by country of birth.[4]	285
29.	Households with chief economic supporter born outside the British Isles: country of birth by persons, by rooms, by sharing of dwelling.[4]	286
30.	Families: household type by persons, by family type.[4]	290
31.	Families: household type by children, by family type.[4]	294
32.	Families: socio-economic group of head by children, by family type.[4]	298
33.	Families: household type by socio-economic group of head.[4]	302
34.	Families: socio-economic group of head by persons, by earners.[4]	303
35.	Families: socio-economic group of head by family type, by earners.[4]	305
36.	Families: persons by family type.[6]	310
37.	Families: household type by persons and children.[6]	318
38.	Families: persons by socio-economic group of head.[6]	333
39.	Families: children's ages in combinations.[4]	339
40.	Families and children: ages of parents by ages of children.[4]	340
41.	Wives and mothers: economic activity and hours worked by number and age of children in family.[4]	341
42.	Married couples: socio-economic group of husband by economic activity of husband and wife.[4]	342
43.	Married couples: husband's socio-economic group, by wife's economic activity, by her age.[4]	343
44.	Married couples: wife's economic activity, by her age, by number of children.[4]	345
45.	Married couples with children in specified age groups: wife's economic activity by her age.[4]	346
46.	Married couples with wife economically active: socio-economic group of husband and wife in combination.[4]	347
AB.	Density (persons per room) by size and type of household, and family type of chief economic supporter, and comparison with hypothetical standards.[4]	348
AD.	Households by family type of chief economic supporter, and rooms occupied, and comparison with hypothetical standards.[4]	352

TABLE TITLES BY VOLUME

CENSUS 1961. ENGLAND AND WALES; INDUSTRY TABLES.

TABLES

Number	Title	Page
	Part I	
1.	Industry (full list) and status.[1]	2
2.	Industry by status.[1]	4
3.	Industry and status by age.[1]	10
4.	Employment status, by economic position, by age.[1]	16
5.	Industry by socio-economic group.[1]	20
6.	Industry by salary/wage earner group (employed persons).[1]	28
7.	Industry by occupation.[1]	32
8.	Industry and status.[2]	122
8A.	Industry and status (persons working in England and Wales enumerated anywhere in Great Britain).[3]	161
9.	Males in part-time employment: present industry by industry in former full-time employment.[1]	163
10.	Males in part-time employment: present status by status in former full-time employment.[1]	164
11.	Residents born outside England and Wales by country of birth or nationality.[1]	166
12.	Residents born outside England and Wales: industry by whether born in the United Kingdom, and whether Commonwealth citizen or alien.[4]	190
12A.	Proportions per 10,000 persons in employment by industry.[5]	196
12B.	Age proportions per 1,000 of each sex, and proportions of females married by industry.[1]	217
	Part II	
13.	Industry by status and occupation by age.[1]	221
14.	Industry by status and occupation.[6]	270

---

[1] England and Wales (10 percent sample).

[2] Regions, conurbations, national and regional urban/rural aggregates, Scotland, Northern Ireland (10 percent sample).

[3] Great Britain, Scotland (10 percent sample).

[4] Regions, conurbations (10 percent sample).

[5] England and Wales, regions, conurbations, national and regional urban/rural aggregates, Scotland, Northern Ireland (10 percent sample).

[6] England and Wales, regions, conurbations (10 percent sample).

GREAT BRITAIN

CENSUS 1961. ENGLAND AND WALES; MIGRATION TABLES.[1]

TABLES

Number	Title	Page
1.	Resident population by sex and duration of residence.[2]	1
2.	Population by sex and duration of residence by age, family status, and socio-economic group.	35
3.	Economically active resident population by sex, by duration of residence, by occupation and industry.	36
4.	Migrants: numbers and proportions.[2]	37
5.	Immigrants by area of former usual residence and sex.[3]	60
5A.	Immigrants by area of former usual residence and sex, by age.[4]	62
6.	Immigrants by area of former usual residence and sex.[5]	74
7.	Immigrants and emigrants by area of usual residence at census and former usual residence.[6]	75
8.	Immigrants and emigrants by families.[6]	180

---

[1] These tables comprise information derived from questions concerning a person's usual address a year before the census date, and the number of years he had lived at his present usual address, put to people included in a 10 percent sample, except members of H.M. forces in camp or on board ship and absent members of private households. Unless otherwise noted, the area for which statistics are given is England and Wales.

[2] England and Wales, regions, conurbations, urban/rural aggregates, counties, local authority areas, conurbation centers, new towns.

[3] Regions, conurbations.

[4] England and Wales, regions, conurbations.

[5] Urban/rural aggregates.

[6] Administrative counties, county boroughs, local authority areas, and new towns with 1,000 or more immigrants or emigrants in sample.

TABLE TITLES BY VOLUME

9.	Immigrants by sex, age, type, and distance of move.[7]	191
10.	Emigrants by sex, age, type, and distance of move.[8]	201
11.	Immigrants by employment moving to rural districts outside conurbations, by sex and workplace.	210
12.	Migrants by sex, age, family status, type and distance of move.	211
13.	Migrants by sex, family status and type of move.[8]	219
14.	Wholly moving private households by socio-economic group of chief economic supporter, tenure, and density of occupation.[9]	224
15.	Wholly moving families enumerated at usual residence by type of move, status of family head, size of family, and persons per room.	262
16.	Migrants by sex, type and distance of move, occupation, industry, and socio-economic group.[7]	264
17.	Immigrants from outside England and Wales by sex, age, family status, and former country of residence.	351
18.	Aliens (including stateless) resident in England and Wales, born outside the United Kingdom, by former country of residence and country of nationality.	358

---

[7] England and Wales, regions, conurbations, urban/rural aggregates.

[8] Regions, conurbations, urban/rural aggregates.

[9] England and Wales, regions, conurbations, urban/rural aggregates, counties, county aggregates, local authority areas, new towns or conurbation centers with 1,000 or more wholly moving households in sample.

GREAT BRITAIN

CENSUS 1961. ENGLAND AND WALES; OCCUPATION, INDUSTRY, SOCIO-ECONOMIC GROUPS.
[58 volumes; see listing on pp. 127-128 above.]

Structure of the Volumes

The tabulations by occupation, industry, and socio-economic groups are published by county in separate volumes. Since the page locations of the tables are not the same in all the volumes, the volumes have been grouped in categories, with the various tables in each volume of each category having the same page locations. The page locations are specified in the Page Location Matrix which follows by category rather than by individual county.

Category I Volumes

Anglesey	Denbighshire	Montgomeryshire
Breconshire	Flintshire	Pembrokeshire
Caernarvonshire	Herefordshire	Radnorshire
Carmarthenshire	Huntingdonshire	Rutland
Cardiganshire	Isle of Ely	Shropshire
Cornwall and the Isles of Scilly	Isle of Wight	Westmorland
	Merionethshire	

Category II Volumes

Devon   Nottinghamshire
Gloucestershire   Worcestershire

Category III Volumes

Cheshire   Hertfordshire
Durham   Lincolnshire
Glamorgan   Warwickshire
Hampshire

Category IV Volumes

Bedfordshire	Leicestershire	Soke of Peterborough
Berkshire	Monmouthshire	Somerset
Buckinghamshire	Norfolk	Suffolk
Cambridgeshire	Northamptonshire	Wiltshire
Cumberland	Northumberland	Yorkshire East Riding
Derbyshire	Oxfordshire	Yorkshire North Riding
Dorset		

Category V Volumes

Kent   Sussex
Staffordshire

Category VI Volumes

Essex   Middlesex

TABLE TITLES BY VOLUME

Category VII Volumes
  Surrey   Yorkshire West Riding

Individual Volumes
  Lancashire   London

TABLES

Number	Title
	Occupation[1]
1.	Occupation and status.
2.	Proportions per 10,000 economically active of each sex by occupation.
	Industry[2]
3.	Industry and status.
4.	Proportions per 10,000 persons in employment by industry.
	Socio-economic group[2]
5.	Socio-economic group (numbers and proportions) of economically active males and economically inactive males (stating an occupation).

PAGE LOCATION MATRIX

Categories of Volumes and Individual Volumes	Pagination by Table Number				
	Table 1	Table 2	Table 3	Table 4	Table 5
Category I	1	3	5	7	10
Category II	1	5	7	12	16
Category III	2	10	16	22	26
Category IV	1	5	8	11	14
Category V	2	10	16	22	28
Category VI	2	18	30	42	48
Category VII	2	14	24	33	40
Lancashire	2	22	38	53	62
London	2	26	44	62	72

[1] Ten percent sample. Unless otherwise noted, the areas for which statistics are given are: county, county boroughs, urban areas with populations of 50,000 or more, county remainders, conurbation centers, new towns.

[2] County, local authority areas, county aggregates, conurbation centers, new towns.

GREAT BRITAIN

CENSUS 1961. ENGLAND AND WALES; OCCUPATION TABLES.[1]

TABLES

Number	Title	Page
1.	Occupation by status.	2
2.	Occupation and status by marital condition and age.	7
3.	Occupation and status by marital condition, by age.	17
4.	Self-employed without employees: occupation by marital condition and age.	41
5.	Self-employed with employees (large establishments): occupation by marital condition and age.	44
6.	Self-employed with employees (small establishments): occupation by marital condition and age.	45
7.	Managers (large establishments): occupation by marital condition and age.	48
8.	Managers (small establishments): occupation by marital condition and age.	50
9.	Foremen and supervisors (manual): occupation by marital condition and age.	52
10.	Foremen and supervisors (non-manual): occupation by marital condition and age.	55
11.	Apprentices, articled clerks and formal trainees: occupation by marital condition and age.	56
12.	Other employees (including professional): occupation by marital condition and age.	58
13.	Persons out of employment (including sick): occupation by marital condition and age.	65
14.	Part-time workers: occupation by marital condition and age.	68
15.	Family workers: occupation by marital condition and age.	71
16.	Institution inmates by occupation.	73
17.	Retired persons: former occupation by marital condition and age.	74
18.	Percentage retired by age (single years), by occupation and socio-economic group.	80
19.	Socio-economic group by marital condition, by age.	118

---

[1] Ten percent sample. Unless otherwise noted, the area for which statistics are given is England and Wales.

TABLE TITLES BY VOLUME

20.	Social class by marital condition, by age.	125
21.	Males in part-time employment: occupation in former full-time employment, by marital condition and age.	126
22.	Males in part-time employment: present occupation by occupation in former full-time employment.	128
23.	Males in part-time employment: present socio-economic group by socio-economic group in former full-time employment.	130
24.	Persons in part-time employment: hours worked by occupation.	131
25.	Persons in part-time employment: hours worked by marital condition, by age.	134
26.	Occupation and status.[2]	136
27.	Socio-economic group and social class.[3]	188
28.	Socio-economic group (numbers and proportions) of economically active males.[2]	198
29.	Socio-economic group (numbers and proportions) of economically inactive males (stating an occupation).[2]	202
30.	Residents born outside England and Wales: occupation by country of birth or nationality.	206
31.	Residents born outside England and Wales: occupation by whether born in the United Kingdom and whether Commonwealth citizen or alien.	210
31A.	Proportions per 10,000 of economically active of each sex, by occupation.[2]	222
31B.	Age proportions and proportion of females married per 1,000 of each sex economically active, by occupation.	261

CENSUS 1961. ENGLAND AND WALES; SOCIO-ECONOMIC GROUP TABLES.[1]

TABLES

Number	Title	Page
1.	Socio-economic group (numbers and proportions) of economically active males.	2

[2] England and Wales, regions, conurbations, national and regional urban/rural aggregates.

[3] Regions, conurbations, urban/rural aggregates.

[1] Ten percent sample.

GREAT BRITAIN

2. Socio-economic group (numbers and proportions) of economically inactive males (stating an occupation).   56

CENSUS 1961. ENGLAND AND WALES; USUAL RESIDENCE TABLES.[1]

TABLES

Number	Title	Page
1.	Comparison of enumerated and resident population.[2]	1
2.	Visitors to England and Wales (persons enumerated in but residing outside England and Wales) by country of residence.[3]	36
3.	Visitors to England and Wales (persons enumerated in but residing outside England and Wales) by country of residence.[4]	40
4.	Visitors to England and Wales by age and marital condition.	46
5.	Visitors to England and Wales by area of residence, age, and marital condition.	46
6.	Visitors to England and Wales by age.[5]	47

CENSUS 1961. ENGLAND AND WALES; WORKPLACE TABLES.[6]

TABLES

Number	Title	Page
1.	Population in employment by areas of residence and workplace.	2
2.	Comparison of resident and day populations.	48

---

[1] No sampling is involved, and, unless otherwise noted, the area for which statistics are given is England and Wales.

[2] England and Wales, regions, conurbations, urban/rural aggregates, local authority areas, conurbation centers, new towns.

[3] England and Wales, regions, conurbations.

[4] Counties, county boroughs, metropolitan boroughs, urban areas with populations of 50,000 or more, county remainders, new towns.

[5] Regions, conurbations, urban/rural aggregates.

[6] Ten percent sample.

3A.	Areas of residence and workplace in combination.[7]	60
3B.	Areas of residence and workplace in combination.[8]	243
3C.	Areas of residence and workplace in combination.[9]	257
4.	Areas of workplace and residence by socio-economic group, occupation, and industry.[10]	259
5.	Workplace movement by socio-economic group, occupation, and industry.[11]	266

CENSUS 1961. GREAT BRITAIN; SCIENTIFIC AND TECHNOLOGICAL QUALIFICATIONS.[1]

TABLES

Number	Title	Page
1.	Qualified persons by subject and type of qualification, by industry.	2
2.	Qualified persons by sex, by occupation.	7
3.	Qualified persons by subject, by occupation.	8
4.	Qualified persons by subject.	11
5.	Qualified persons by occupation, sex, age (and marital condition for females).	12
6.	Qualified persons by subject, sex, and age.	13

---

[7] Local authority areas.

[8] Conurbations, conurbation centers, conurbation remainders.

[9] New towns.

[10] Conurbations.

[11] Local authority areas where sample figure of inward or outward movement is 10,000 or more.

[1] Ten percent sample. Information was obtained on the <u>type of qualification</u> (university degree, graduate or corporate membership in a professional association, etc.) for persons qualified in science and technology, excluding medicine, dentistry, pharmacy, optics, veterinary science, architecture, economics, geography, and the social sciences. The <u>subject of qualification</u> (biology, chemistry, etc.) was also obtained. The area for which statistics are given is England, Wales, and Scotland.

GREAT BRITAIN

7.	Qualified persons by occupation, by sex and age (and by marital condition for females).	14
8.	Qualified persons working in the United Kingdom who were born outside the British Isles.	18

CENSUS 1961. GREAT BRITAIN; SUMMARY TABLES.[1]

TABLES

Number	Title	Page
	Population	
1.	Population by sex and intercensal changes, 1801-1961.	1
2.	Estimated population by sex as at the middle of each year, 1801-1961.	2
3.	Number of areas by type and size of population.	3
	Age and Marital Condition	
4.	Age (single years) and marital condition.	4
5.	Age: numbers and successive censuses, 1841-1961 (thousands).	6
6.	Age and marital condition at successive censuses, 1851-1961 (thousands).	7
	Non-private Households	
7.	Population enumerated in private and non-private households.	9
8.	Population enumerated in and outside private households by sex and age.	9
	Birthplace and Nationality	
9.	Birthplaces, nationalities, and citizenships of residents of Great Britain born outside the British Isles.	10
10.	Nationalities and citizenships of residents of Great Britain born outside the British Isles or with birthplace not stated.	12

---

[1] Unless otherwise noted, no sampling is involved. The area for which statistics are given is England, Wales, and Scotland.

Fertility

11(i).	Married women by age at census and size of family.	13
11(ii).	Widowed and divorced women by age at census and size of family.	14
12(i).	Women with uninterrupted first marriage: size of family by duration of marriage and age at marriage.	15
12(ii).	Current fertility of women with uninterrupted first marriage: age at marriage by duration of marriage and size of family.	18
13.	Women with uninterrupted first marriage: size of family by year of marriage and age at marriage.	19

Buildings, Dwellings, and Households

14.	Dwellings by building type, rooms and household spaces.	26
15.	Private households by size, rooms occupied, sharing of dwellings and type of building.	28
16.	Private households by type of building (proportions per 1,000).	33
17.	Private households by density of occupation (persons per room).	33
18.	Population in private households by density of occupation (persons per room).	34

Tenure and Household Arrangements

19.	Dwellings by tenure, household spaces and rooms.	35
20.	Private households by size, rooms occupied, sharing of dwellings, and tenure.	38
21.	Private households by tenure (proportions per 1,000) and proportions of households sharing a dwelling in each tenure category (percent).	44
22.	Dwellings by availability of certain household arrangements.	45
23.	Private households by availability of certain household arrangements.	46
24.	Private households sharing or lacking household arrangements: all combinations.	47

Migration[2]

25.	Resident population by sex and duration of residence by age, family status, and socio-economic group.	48

---

[2]Ten percent sample.

26.	Economically active resident population by sex, by duration of residence, by occupation and industry.	49
27.	Migrants by sex, age, family status, type and distance of move.	50
28.	Wholly moving private households by socio-economic group of chief economic supporter, tenure, and density of occupation.	58
29.	Migrants by sex, type and distance of move, occupation, industry, and socio-economic group.	59
30.	Immigrants from outside Great Britain by sex, age, family status, and former country of residence.	63

Occupation^2

31.	Occupation by status.	64
32.	Occupation and status by marital condition and age.	70
33.	Occupation and status by marital condition, by age.	83
34.	Persons in part-time employment: hours worked by occupation.	107
35.	Persons in part-time employment: hours worked by marital condition, by age.	108

Industry^2

36.	Industry (full list) and status.	110
37.	Industry and status by age.	114

Terminal Education Age

38.	Population aged 15 and over in 9 age sections classified by 16 terminal education age groups.	120

Household Composition

39.	Households: socio-economic group of chief economic supporter by tenure, by persons, by rooms.	122
40.	Households: household type by persons.	143
41.	Households: family type of chief economic supporter by his economic activity and household type.	144
42.	Households containing persons of pensionable age: household type, persons of pensionable age in households: household type by sex, age, marital condition, and economic activity.	146
43.	One- and two-person households containing persons of pensionable age: household type by rooms, by sharing of dwelling.	147
44.	Husbands' and wives' ages in combination by socio-economic group of husband.	148

TABLE TITLES BY VOLUME

45.	Households: persons by earners by children.	155
46.	Wives and mothers: economic activity and hours worked by number and age of children in family.	157
47.	Married couples: socio-economic group of husband by economic activity of husband and wife.	158

CENSUS 1961. REPORT ON JERSEY, GUERNSEY AND ADJACENT ISLANDS.[1]

TABLES

Number	Title	Page
	Population	
1.	Population 1821-1961 and intercensal variations.	1
1A.	Population of the various islands at each census, 1821-1961.[2]	1
2.	Acreage, population 1931-1961, and intercensal variations, private households and dwellings 1961.[3]	2
2A.	Populations of vingtaines in each civil parish.[4]	2
	Usual Residence	
3.	Visitors to the Channel Islands by country of residence.	3
	Sex, Age, and Marital Condition	
4.	Age (single years).	4
5.	Age and marital condition.[5]	5
	Birthplace and Nationality	
6.	Birthplaces and nationalities of the whole population.	7
7.	Birthplaces, nationalities, and citizenships of residents of the Channel Islands born outside the British Isles.	8

---

[1] Unless otherwise noted, the areas for which statistics are given are Jersey, Guernsey, and adjacent islands. No sampling is involved.

[2] Guernsey and adjacent islands.

[3] Jersey, Guernsey and adjacent islands; civil parishes.

[4] Jersey.

[5] Jersey, St. Helier civil parish, Guernsey and adjacent islands, Alderney, Sark, St. Peter Port civil parish.

GREAT BRITAIN

Housing and Households

8.	Dwellings by rooms and household spaces.[3]	9
9.	Private households by size, rooms occupied, and sharing of dwellings.[3]	17
10.	Dwellings by tenure and rooms.	28
11.	Private households by size, rooms occupied, sharing of dwelling and tenure.	30
12.	Private households by availability of certain household arrangements.[3]	34
12A.	Private households with accommodation for paying guests: with in-laws usually resident.[6]	38
12B.	Households in dwellings connected by public sewer.[6]	38

Non-private Households

13.	Non-private households (excluding hotels) by type, population, and inmates.[3]	39
14.	Hotels and boarding houses: number, rooms, and population.	40

Old People

15.	One- and two-person households containing persons of pensionable age by sex/condition combination.[3]	41

Migration

16.	Resident population by sex and duration of residence.[3]	42
16A.	Heads of households not born in Jersey by length of residence.[6]	43
17.	Immigrants by area of usual residence at census and former usual residence, sex, and age.	44
18.	Economically active immigrants: occupation, socio-economic group, and industry by area of former usual residence and sex.	45

Terminal Education Age

19.	Population over school leaving age in 4 age sections by 6 terminal education ages.[7]	47
20.	Qualified persons by subject and type of qualification, by occupation and industry.	48

---

[6] Jersey, civil parishes.

[7] Jersey, Guernsey and adjacent islands, St. Helier civil parish, St. Peter Port civil parish.

Language

21.	Population aged 15 and over, speaking English or French and able to understand written French.[8]	50

Occupation and Industry

22.	Status by sex.[7]	51
23.	Occupation by status, marital condition, and age.[7]	52
24.	Socio-economic group of economically active and retired males.[7]	70
25.	Part-time workers: age and occupation by hours worked.	71
26.	Industry by status.[7]	73

Workplace

27.	Population in employment, by area of residence and workplace.[9]	79

Fertility

28.	Married women and widowed and divorced women, by age and size of family.	80

CENSUS 1961. REPORT ON ISLE OF MAN. PART I. POPULATION AND HOUSING.[1]

TABLES

Number	Title	Page
	Population	
1.	Population 1821-1961 and intercensal variations.	1
2.	Acreage, population 1931-1961, and intercensal variations, private households, and dwellings 1961.[2]	2
3.	Population 1931-1961 and intercensal variations, private households, and dwellings 1961.[3]	3

---

[8] Jersey, St. Helier civil parish.

[9] Jersey, civil parishes, Guernsey and adjacent islands, Alderney, Sark, Herm and Jethou, Guernsey remainder.

[1] Unless otherwise noted, the area for which statistics are given is the Isle of Man. No sampling is involved.

[2] Isle of Man, civil parishes.

[3] Town and village districts with their constituent civil parishes and wards.

GREAT BRITAIN

Sex, Age, and Marital Condition

4. Age and marital condition.[4]   4

Birthplace and Nationality

5. Birthplaces and nationalities of the whole population.   6
6. Birthplaces, nationalities, and citizenships of residents of the Isle of Man born outside the British Isles.   7

Housing and Households

7. Dwellings by rooms and household spaces.[5]   8
8. Private households by size, rooms occupied, and sharing of dwellings.[5]   12
9. Dwellings by tenure and rooms.   17
10. Private households by size, rooms occupied, sharing of dwelling, and tenure.   18
11. Private households by availability of certain household arrangements.[5]   20

Non-private Households

12. Institutions.[6]   27

Old People

13. One- and two-person households containing persons of pensionable age by sex/condition combination.[5]   28

CENSUS 1961. REPORT ON ISLE OF MAN. PART II. MIGRATION, ECONOMIC ACTIVITY AND OTHER TOPICS.[1]

TABLES

Number	Title	Page
	Usual Residence	
14.	Visitors to the Isle of Man by country of residence.	29

---

[4] Isle of Man, town districts, and island remainder.

[5] Isle of Man, town districts, village districts, and aggregate of parish districts.

[6] Isle of Man, town and village districts.

[1] Unless otherwise noted, the area for which statistics are given is the Isle of Man. No sampling is involved.

Sex and Age
15. Age--single years. ... 30

Non-private Households
16. Hotels and boarding houses: number, rooms, and population.[2] ... 31
16A. Accommodation for visitors.[3] ... 31

Migration
17. Resident population by duration of residence and sex.[2] ... 32
18. Immigrants by area of usual residence at census and former usual residence, sex, and age. ... 33
19. Economically active immigrants: socio-economic group by area of former usual residence and sex. ... 33

Terminal Education Age
20. Population over school leaving age in 4 age sections by 6 terminal education ages.[4] ... 34
21. Qualified persons by subject and type of qualification, by occupation and industry. ... 36

Language
22. Persons aged 3 and over speaking both English and Manx.[5] ... 37

Occupation and Industry
23. Status by sex. ... 38
24. Social class by economically active and retired males. ... 38
25. Occupation by status, marital condition, and age. ... 39
26. Industry by status. ... 44

Workplace
27. Areas of residence and workplace in combination.[6] ... 46

---

[2] Isle of Man, town districts, village districts, and aggregate of parish districts.

[3] Isle of Man, town districts, village districts, and parish districts.

[4] Isle of Man, town districts, Onchan village district, and remainder of village districts.

[5] Isle of Man, Douglas.

[6] Town districts, Onchan village district.

GREAT BRITAIN

Fertility

28. Married women and widowed and divorced women by age and size of family.   47

CENSUS 1961. WALES (INCLUDING MONMOUTHSHIRE); REPORT ON WELSH SPEAKING POPULATION.

TABLES

Number	Title	Page
1.	Welsh speaking population (aged 3 and over).[1]	1
2.	Welsh speaking population (aged 3 and over) by ages.[2]	22
3.	Welsh speaking population: percentage of population aged 3 and over at successive censuses, 1911-1961.[3]	33
4.	Welsh speaking population: percentage in each age-group at successive censuses, 1911-1961.[4]	33

---

[1] Wales, counties, local authority areas, wards, civil parishes in rural districts, Cwmbran New Town.

[2] Wales, counties, county boroughs, areas with populations of 20,000 or more, county remainders, Cwmbran New Town.

[3] Wales, counties, county boroughs, urban areas with populations of 50,000 or more.

[4] Wales.

G R E E C E

MAIN ENTRIES AND VOLUME NOTES:[1]

Greece. Ethnike Statistike Hyperesia.
 Population de la Grèce au recensement du 19 mars 1961; population de fait par départements, éparchies, communes-dèmes, communes, et localités. Athènes, 1962.
  183 p. 30cm. (NUC63-14169)

Greece. Ethnike Statistike Hyperesia.
 Résultats du recensement de la population et des habitations effectué le 19 mars 1961. Athènes, 1963-1968.
  3 v. 35cm.

   Vol. I. Population par subdivisions géographiques et administratives. 1964. 618 p.

   Vol. II. Caractéristiques démographiques, sociales et économiques de la population. Conditions de logement des ménages. Données par communes-dèmes et communes. Données par départements, avec subdivision de chaque département en circonscriptions urbaines, semi-urbaines, et rurales. 10 fascicules.[2]

   1. La Grèce entière, région d'Athènes, agglomérations urbaines et villes de 50,000 habitants et au-dessus.
   2. Le reste de la Grèce centrale et Eubée.
   3. Péloponèse.
   4. Îles Ioniennes.
   5. Épire.
   6. Thessalie.
   7. Macédoine.
   8. Thrace.
   9. Îles de la Mer Égée.
   10. Crète.

---

[1] The titles and texts for four of the volumes in this census are in Greek and French; the six remaining volumes are in Greek and English. The French and English versions are used below.

[2] The total pagination of each fascicle is not given because the pages are not consecutively numbered.

Vol. III. Caractéristiques démographiques, sociales et économiques de la population. Conditions de logement des ménages. Grèce entière; régions géographiques; circonscriptions urbaines; semi-urbaines et rurales. 1968. vii, 301 p.

Greece. Ethnike Statistike Hyperesia.
Results of the population and housing census of 19 March 1961. Sample elaboration. Athens, 1962-
v. 33cm.

Vol. I. Demographic characteristics. 1962. 31 p.

Vol. II. Education (general education level, professional training). 1962. 31 p.

Vol. III. Employment (economically active population, employed and unemployed, hours of work). 1962. 63 p.

Vol. IV. Economically inactive population. 1962. 29 p.

Vol. V. Internal migration. 1963. 54 p.

Vol. VI. Housing conditions. 1963. 31 p. [Excluded.]

Vol. VII. Composition and other characteristics of households. [Announced but not yet published.]

ENGLISH TRANSLATION OF CENSUS AND VOLUME TITLES:

Population of Greece, census of March 19, 1961; de facto population by department, district, urban municipality, rural municipality, and locality.

Results of the census of population and housing on March 19, 1961.

I. Population by geographic and administrative subdivision.

II. Demographic, social, and economic characteristics of the population. Housing conditions of households. Data by municipality and commune. Data by department with subdivision of each department into urban, semi-urban, and rural areas. [10 fascicles]

III. Demographic, social, and economic characteristics of the population. Housing, households. All Greece; geographical regions; urban, semi-urban, and rural areas.

Results of the population and housing census of 19 March 1961; sample elaboration.

I. Demographic characteristics.

II. Education.

III. Employment.

IV. Economically inactive population.

GLOSSARY

   V. Internal migration.

  VI. Housing conditions.

 VII. Composition and other characteristics of households.

ENUMERATION DATE: March 19, 1961.

ENUMERATION BASIS: De facto; de jure.

GLOSSARY

ADMINISTRATIVE, GEOGRAPHICAL, AND POLITICAL UNITS

Level	English Translation[1]	French Term
National	Kingdom of Greece	Royaume de Grèce
Intermediate	Department	Département
Local	District	Éparchie
	Urban municipality	Commune-dème
	Rural municipality (commune)	Commune

Population cluster ("locality") concepts:[2]

	Locality	Localité
	Rural, semi-urban, and urban areas	Circonscriptions rurales, semi-urbaines, urbaines

Other concept:

	Geographical region	Région géographique

    In Greece, <u>departments</u> are administrative divisions which correspond roughly to states or provinces of other nations. They are composed of <u>districts</u>, which serve as voting jurisdictions and sometimes as local authority units in remote areas.

    Districts are made up of <u>municipalities</u>, of which there are two kinds: <u>urban municipalities</u> and <u>rural municipalities</u> (or communes).

---

[1] Terminology is that used in the English version of the "Sample elaboration" volumes of the census.

[2] For a discussion of this concept, see <u>European Population Censuses: The 1960 Series</u>, U.N. Doc.: ST/CES/3 (1964), p. 108.

Urban municipalities have more than 10,000 inhabitants and serve as capitals of departments.

Rural municipalities (or communes) consist of a village or a settlement, or groups of villages or settlements, with more than 5,000 inhabitants and an elementary school. They must have adequate revenue to support local government activities.

Localities are population agglomerations where 10 or more houses normally shelter at least 50 persons. They are separated by an empty stretch of at least 200 meters or by commune boundaries from other similar agglomerations or institutions housing 50 or more persons.

Rural, semi-urban, and urban areas are derived from classifying urban and rural municipalities by the size of their largest population center as follows:

Rural areas are urban and rural municipalities whose largest population center has fewer than 2,000 inhabitants.

Semi-urban areas are urban and rural municipalities whose largest population center has 2,000 or more but fewer than 10,000 inhabitants.

Urban areas are urban and rural municipalities whose largest population center has at least 10,000 inhabitants.

Geographical regions refer to the traditional regional divisions: Athens, Central Greece and Euboea, the Peloponnesus, the Ionian Islands, Epirus, Thessaly, Macedonia, Thrace, the Aegean Islands, and Crete.

TABLE OF DEMOGRAPHIC CONCEPTS

The demographic concepts which appear in the following table correspond to the recommendations of the European Conference of Statisticians, unless otherwise noted. They are defined and discussed in the appendixes and in footnotes. Terms in parentheses are alternative forms which appear in the table titles. Terms in brackets are not used in the table titles, but they are included here in order to give logical completeness to the various classifications in the table.

Appendix	English Table Term	French Table Term
	ECONOMIC CHARACTERISTICS	
A	Economic activity	Activité économique
	Economically active[1]	Personnes actives
	Economically inactive	Personnes non-actives

---

[1] Economically active persons are those working regularly or

TABLE TITLES BY VOLUME

B	Employment status	Statut dans la profession
C	[Industrial classification]	...
	Division(s)[2]	Branche d'activité économique
	[Major group]	...
	Group(s)	Groupe des branches d'activité économique
D	[Occupational classification]	...
	[Major group]	
	[Minor group]	...
	Unit group(s)	Groupe des professions individuelles

HOUSEHOLD AND FAMILY CHARACTERISTICS[3]

F	[Household][4]	...,
	Private household	Ménage
	Institution	Ménage collectif

TABLE TITLES BY VOLUME

POPULATION OF GREECE, CENSUS OF MARCH 19, 1961;
DE FACTO POPULATION BY DEPARTMENT, DISTRICT, URBAN
MUNICIPALITY, RURAL MUNICIPALITY, AND LOCALITY

TABLES

Number	Title	Page
	A. Summary Tables	
I.	De facto population by geographical region.	13
II.	De facto population by department.	14

---

working at least ten hours during the week prior to census date, and persons seeking work (employed and unemployed).

[2]The term "division" may also designate an abbreviated industrial classification of the following categories: (1) agriculture, forestry, fishing; (2) mining, manufacturing, construction, etc.; (3) commerce, transportation, communication, etc.; (4) not determined or not declared.

[3]Tabulations of family characteristics are not given, but they can be derived from the existing tables.

[4]The classification by household employs a national definition based on the housekeeping unit concept.

III.	De facto population by district.		16
	B. Analytical Tables		
IV.	De facto population by urban municipality, rural municipality (commune), and locality.		25
	Department of	Aetolia and Acarnania	25
	Districts:	Valtos	25
		Vonitsa and Xiromeron	26
		Missolonghi	27
		Navpaktos	28
		Trichonis	29
	Department of	Argolis	31
	Districts:	Argos	31
		Ermioni	32
		Nauplia	32
	Department of	Arcadia	33
	Districts:	Gortinia	33
		Kinouria	35
		Mantinea	36
		Megalopolis	38
	Department of	Arta	39
	District:	Arta	39
	Department of	Attica	42
	Districts:	Aegina	42
		Attica	42
		1. Athens, region	42
		2. Remainder of eparchy	42
		Kithira	44
		Megara	45
		Troizin	45
		Idhra	46
	Department of	Akhaia	46
	Districts:	Aigaleos	46
		Kalavrita	47
		Patrai	49
	Department of	Voiotia	52
	Districts:	Thebes	52
		Levadhia	53
	Department of	Drama	54
	District:	Drama	54
	Department of	Dodecanesos	56
	Districts:	Kalimnos	56
		Karpathos	56
		Kos	56
		Rhodes	57

TABLE TITLES BY VOLUME

Department of	Evros	58
Districts:	Alexandroupolis	58
	Dhidhimotikhon	59
	Orestias	59
	Samothrace	60
	Souphlion	60
Department of	Evvoia	61
Districts:	Istiaia	61
	Karystos	61
	Chalkis	63
Department of	Eurytania	65
District:	Eurytania	65
Department of	Zakynthos (Zante)	67
District:	Zakynthos (Zante)	67
Department of	Ilia	68
Districts:	Ilia	68
	Olympia	72
Department of	Ematheia (Hematheia)	73
Districts:	Ematheia (Hematheia)	73
	Naoussa	74
Department of	Iraklion	75
Districts:	Viannos	75
	Kainourion	75
	Malezivi	76
	Monophatsi	77
	Pedhias	78
	Pyrghiotissa	79
	Temenos	80
Department of	Thesprotia	80
Districts:	Thiamis	80
	Margariti	81
	Souli (Paramythia)	81
	Philiates	82
Department of	Salonika	83
Districts:	Salonika	83
	Langadhas	84
Department of	Jannina	86
Districts:	Dodona	86
	Konitsa	91
	Metsovon	92
	Pogonion	92
Department of	Kavala	93
Districts:	Thasos	93
	Kavala	93
	Nestos	94
	Pangaion	94

GREECE

Department of Kardhitsa	95
District: Kardhitsa	95
Department of Kastoria	98
District: Kastoria	98
Department of Korfu	100
Districts: Korfu	100
Paxos	103
Department of Cephalonia	103
Districts: Ithaca	103
Kranea	103
Palios	104
Sami	105
Department of Kilkis	106
Districts: Kilkis	106
Paiania	108
Department of Kozani	108
Districts: Voion	108
Grevena	110
Eordaia	111
Kozani	112
Department of Korinthia	114
District: Korinthia	114
Department of Cyclades	116
Districts: Andros	116
Thira	117
Ceos (Kea)	118
Milo	118
Naxos	119
Paros	119
Syra	120
Tinos	120
Department of Lakonia	121
Districts: Yithion	121
Epidaurus Limera	121
Lakedaimon	123
Oitilon	125
Department of Larisa	125
Districts: Agia	125
Elasson	126
Larisa	127
Tirnavos	128
Farsala	128
Department of Lasithi	129
Districts: Ierapetra	129
Lasithi	130
Mirabella	130
Sitia	131

TABLE TITLES BY VOLUME

Department of	Lesbos	133
Districts:	Lemnos (Limnos)	133
	Methymna	133
	Mytilene	134
	Plomarion	135
Department of	Levkas	135
District:	Levkas	135
Department of	Magnisia	136
Districts:	Almiros	136
	Volo	137
	Skopelos	138
Department of	Messinia	139
Districts:	Kalamato	139
	Messini	140
	Pilias	142
	Trifillion	144
Department of	Xanthi	146
District:	Xanthi	146
Department of	Pella	147
Districts:	Almopia	147
	Giannitsa	148
	Edessa	149
Department of	Pieria	150
District:	Pieria	150
Department of	Preveza	151
District:	Nicopolis and Parga	151
Department of	Rethimni	153
Districts:	Ayios Vasilios	153
	Amari	153
	Mylopotamos	154
	Rethimni	155
Department of	Rodhopi	156
Districts:	Komotini	156
	Sapai	157
Department of	Samos	158
Districts:	Ikaria	158
	Samos	159
Department of	Serres	160
Districts:	Bisaltia	160
	Serres	160
	Sintica	161
	Phyllis	162
Department of	Trikala	163
Districts:	Kalambaka	163
	Trikala	164

GREECE

Department of	Fthiotis	166
Districts:	Dhomokos	166
	Locris	167
	Fthiotis	168
Department of	Florina	170
District:	Florina	170
Department of	Fokis	172
Districts:	Doris	172
	Parnassus	173
Department of	Khalkidiki	174
Districts:	Arnaia	174
	Khalkidiki	175
Department of	Canea	176
Districts:	Apokoronas	176
	Kissamos	177
	Kidhonia	179
	Selinous	180
	Sphakia	181
Department of	Chios	182
Districts:	Chios	182
	Aghion Oros	183

RESULTS OF THE CENSUS OF POPULATION AND HOUSING
ON MARCH 19, 1961

VOLUME I. POPULATION BY GEOGRAPHIC AND ADMINISTRATIVE SUBDIVISION.

TABLES

Number	Title	Page
1.	Enumerated (de facto) population of each urban municipality, commune, and locality divided into (a) centralized and dispersed population; (b) (1) inhabitants of the urban municipality or commune registered in the Municipal Registers of their own community; (2) inhabitants registered in the Registers of another community; (3) foreigners. Resident (de jure) population of each municipality and commune, divided into (1) persons present on the day of the census in the urban municipality or commune and registered in Municipal Registers; (2) persons counted in urban municipalities or communes other than those where they were registered; (3) persons registered in the Municipal Registers but absent, outside of Greece.	
	A. Data by urban municipality, commune, and locality.	1

	B. Summary data: (a) by district, p. 327; (b) by department, p. 331; (c) by geographical region, all of Greece, p. 332.	327
2.	Localities and population of localities, grouped by order of importance: (a) by district, (b) by department, (c) by geographical region, all of Greece.	335
3.	Urban and rural municipalities (communes) by urban, semi-urban, and rural areas. Number of inhabitants, area, and density of the population of these municipalities.	
	A. Data by urban municipality and commune.	359
	B. Summary data: (a) by district, p. 411; (b) by department, p. 423; (c) by geographical region, all of Greece, p. 427.	411
4.	Altitude of each locality and weighted averages of the altitudes.[1]	
	A. Altitude of each locality and weighted average of the altitude of each urban municipality or commune.	431
	B. Weighted averages of the altitudes by district, department, and geographical region, all of Greece: (a) by district, p. 586; (b) by department, p. 588; (c) by geographical region, all of Greece, p. 589.	586
5.	Localities and total population of these localities, by zone of altitude: (a) by district, p. 593; (b) by department, p. 612; (c) by geographical region, all of Greece, p. 618.	593

VOLUME II. DEMOGRAPHIC, SOCIAL, AND ECONOMIC CHARACTERISTICS OF THE POPULATION. HOUSING CONDITIONS OF HOUSEHOLDS. DATA BY MUNICIPALITY AND COMMUNE. DATA BY DEPARTMENT WITH SUBDIVISION OF EACH DEPARTMENT INTO URBAN, SEMI-URBAN, AND RURAL AREAS. [10 FASCICLES]

Structure of the Volume

The table numbers and table titles for all the fascicles of Volume II are identical. Page locations, however, vary from fascicle to fascicle. Hence, it is necessary to consult the

---

[1] Altitude of each locality, weighted average of the altitude of each city or town, and weighted averages of the altitudes by district, department, and geographical region, for all of Greece.

GREECE

Tables list as well as the Page Location Matrixes which follow in order to locate desired data.

The procedure for obtaining page locations of specific data involves two steps:
1. Consult the Tables list for desired data and table number.
2. Consult the individual Page Location Matrix for the fascicle desired. Each fascicle has the same six tables, for the region as a whole and for each department, which are separately paged.

TABLES

Number | Title

1. Demographic and social characteristics (sex, age, marital status, educational status, and employment status) of the population.

2. Demographic and social characteristics (sex, age, marital status, educational status, and employment status) of the population by urban, semi-urban, and rural areas.

3. Economically active population, employed and unemployed, by sex and occupational group. Total area, and urban, semi-urban, and rural areas.

4. Private households by housing conditions, showing number of artisans residing in the household.

5. Private households by number of members and households having a resident artisan. Total area, and urban, semi-urban, and rural areas.

6. Private households living in regular dwellings by number of members and number of rooms available.

TABLE TITLES BY VOLUME

PAGE LOCATION MATRIXES

Fascicle 1. All of Greece, Greater Athens, urban agglomerations and cities of 50,000 inhabitants and more.

Tables	Regional Tables	Region of Athens	Urban Agglomerations	Cities with 50,000 or more inhabitants
1	9	25	45	--
2	10	29	50	91
3	14	31	56	97
4	20	38	76	--
5	21	40	79	118
6	22	42	85	124

Fascicle 2. Remainder of Central Greece and Evvoia.

Tables	Regional Tables	Department of Aetolia and Acarnania	Department of Attica	Department of Voiotia	Department of Evvoia	Department of Eurytania	Department of Fthiotis	Department of Fokis
1	2.1	2.1.1	2.2.1	2.3.1	2.4.1	2.5.1	2.6.1	2.7.1
2	.4	.15	.7	.6	.11	.7	.12	.7
3	.6	.17	.9	.8	.13	.9	.14	.9
4	.13	.23	.16	.14	.20	.14	.20	.15
5	.15	.31	.20	.17	.25	.17	.26	.18
6	.17	.33	.22	.19	.27	.19	.28	.20

Fascicle 3. Peloponnesus.

Tables	Regional Tables	Department of Argolis	Department of Arcadia	Department of Akhaia	Department of Ilia	Department of Korinthia	Department of Lakonia	Department of Messinia
1	3.1	3.1.1	3.2.1	3.3.1	3.4.1	3.5.1	3.6.1	3.7.1
2	.4	.6	.16	.15	.14	.9	.11	.18
3	.6	.8	.18	.17	.16	.11	.13	.20
4	.13	.14	.25	.24	.22	.17	.19	.26
5	.15	.17	.33	.33	.29	.21	.25	.35
6	.17	.19	.35	.34	.31	.23	.27	.37

Fascicle 4. Ionic Islands.

Tables	Regional Tables	Department of Zakynthos (Zante)	Department of Korfu	Department of Cephalonia	Department of Levkas
1	4.1	4.1.1	4.2.1	4.3.1	4.4.1
2	.3	.4	.8	.7	.4
3	.5	.6	.10	.9	.6
4	.12	.12	.16	.15	.12
5	.13	.14	.20	.19	.14
6	.15	.16	.22	.21	.16

Fascicle 5. Epirus.

Tables	Regional Tables	Department of Arta	Department of Thesprotia	Department of Jannina	Department of Preveza
1	5.1	5.1.1	5.2.1	5.3.1	5.4.1
2	.3	.6	.8	.20	.6
3	.5	.8	.10	.22	.8
4	.12	.14	.16	.28	.14
5	.13	.17	.20	.38	.17
6	.15	.19	.22	.40	.19

Fascicle 6. Thessaly.

Tables	Regional Tables	Department of Kardhitsa	Department of Larisa	Department of Magnisia	Department of Trikala
1	6.1	6.1.1	6.2.1	6.3.1	6.4.1
2	.3	.10	.11	.7	.9
3	.5	.12	.13	.9	.11
4	.12	.18	.20	.16	.17
5	.13	.23	.26	.19	.22
6	.15	.25	.28	.21	.24

Fascicle 7. Macedonia.

Tables	Regional Tables	Department of Drama	Department of Ematheia (Hematheia)	Department of Salonika	Department of Kavala	Department of Kastoria	Department of Kilkis
1	7.1	7.1.1	7.2.1	7.3.1	7.4.1	7.5.1	7.6.1
2	.5	.6	.5	.9	.7	.5	.6
3	.7	.8	.7	.11	.9	.7	.8
4	.14	.15	.13	.18	.16	.13	.14
5	.17	.18	.16	.22	.19	.15	.17
6	.19	.20	.18	.24	.21	.17	.19

Fascicle 7. Macedonia (continued)

Tables	Department of Kozani	Department of Pella	Department of Pieria	Department of Serres	Department of Florina	Department of Khalkidiki	Mount Athos
1	7.7.1	7.8.1	7.9.1	7.10.1	7.11.1	7.12.1	7.1
2	.15	.7	.4	.10	.7	.6	.2
3	.17	.9	.6	.12	.9	.8	.4
4	.24	.15	.12	.18	.15	.14	.8
5	.32	.18	.14	.23	.18	.17	.9
6	.34	.20	.16	.25	.20	.19	.11

Fascicle 8. Thrace.

Tables	Regional Tables	Department of Evros	Department of Xanthi	Department of Rodhopi
1	8.1	8.1.1	8.2.1	8.3.1
2	.2	.7	.4	.4
3	.4	.9	.6	.6
4	.11	.15	.12	.12
5	.12	.18	.14	.14
6	.14	.20	.16	.16

Fascicle 9. Islands of the Aegean Sea.

Tables	Regional Tables	Department of Dodecanesos	Department of Cyclades	Department of Lesbos	Department of Samos	Department of Chios
1	9.1	9.1.1	9.2.1	9.3.1	9.4.1	9.5.1
2	.3	.6	.10	.8	.5	.5
3	.5	.8	.12	.10	.7	.7
4	.12	.14	.18	.16	.13	.13
5	.14	.17	.23	.20	.15	.15
6	.16	.19	.25	.22	.17	.17

GREECE

Fascicle 10. Crete.

Tables	Regional Tables	Department of Iraklion	Department of Lasithi	Department of Rethimni	Department of Canea
1	10.1	10.1.1	10.2.1	10.3.1	10.4.1
2	.3	.13	.7	.10	.11
3	.5	.15	.9	.12	.13
4	.12	.21	.15	.18	.19
5	.14	.28	.19	.23	.25
6	.16	.30	.21	.25	.27

VOLUME III. DEMOGRAPHIC, SOCIAL, AND ECONOMIC CHARACTERISTICS OF THE POPULATION. HOUSING, HOUSEHOLDS. ALL GREECE; GEOGRAPHICAL REGIONS; URBAN, SEMI-URBAN, AND RURAL AREAS.

TABLES

Number	Title	Page
1.	Population by sex and age.	2
2.	Population by marital status, sex, and age group.	44
3.	Population by nationality and sex.	72
4.	Population 10 years of age and older by level of education, sex, and age group.	80
5.	Graduates of institutions of higher education by kind of degree and sex.	102
6.	Persons having a license, certificate, or diploma from a technical school or professional apprenticeship (of the first or second degree) by kind of title and sex.	104

TABLE TITLES BY VOLUME

7.	Population 10 years of age and older, grouped according to those who are economically active (employed and unemployed) and economically inactive, by sex and age group.	108
7a.	Population 10 years of age and older, grouped according to those who are economically active (employed and unemployed) and economically inactive, by sex and age group. [Continuation of Table 7.]	110
8.	Economically active persons, including unemployed, by industrial division and sex.	122
9.	Economically active persons by industrial division, sex, and age group.	124
10.	Economically active persons by industrial division, abbreviated major group of occupations, and sex.	150
11.	Economically active persons by industrial division, employment status, and sex.	176
12.	Economically active persons by industrial division, level of education, and sex.	178
13.	Economically active persons by industrial group and sex.	192
14.	Economically active persons by major group of occupations, sex, and age group.	206
15.	Economically active persons by employment status, sex, and age group.	232
16.	Economically active persons by major group of occupations, employment status, and sex.	234
17.	Economically active persons by individual occupation and sex.	238
18.	Economically inactive persons by reason for not working, sex, and age group.	254
19.	Economically active persons by principal source of income, sex, and age group.	268
20.	Division of the population into private households and institutional households by sex and age group.	282
21.	Private households by number of members living in "normal" dwellings and "rudimentary-mobile-improvised" dwellings.	284
22.	Private households by the number of members, and the sex and marital status of the head of the household.	285
23.	Private households by the number of members and the occupation of the head of the household.	292

GREECE

24.	Private households by the number of members, and the age and employment status of the head of the household.	294
25.	Institutional households by type and number of members, divided into staff, inmates, and boarders.	296

RESULTS OF THE POPULATION AND HOUSING CENSUS
OF 19 MARCH 1961. SAMPLE ELABORATION

VOLUME I. DEMOGRAPHIC CHARACTERISTICS.

TABLES

Number	Title	Page
1.	Total population by sex and type of household.	19
2.	Total population by sex and age.	20
3.	Greek population 15 years of age and over, by age group, sex, and marital status.	21
4.	Total population and population in private households by sex. Total population in urban-rural areas by sex.	23
5.	Total population by sex and broad age groups.	26
6.	Total population aged 15 years and over, by sex and marital status.	29
7.	Foreign nationals by sex and nationality.	31

VOLUME II. EDUCATION (GENERAL EDUCATION LEVEL, PROFESSIONAL TRAINING).[1]

TABLES

Number	Title	Page
1.	Total population 10 years of age and over, by sex, age, and general level of education and the proportion illiterate.	15

---

[1] All persons 10 years of age and over were classified according to the highest level of general education they had completed: higher, secondary, or primary. Completing "higher" education means receiving a diploma from an educational institution higher than a *gymnasium*; completing "secondary" education, a diploma from a *gymnasium*; completing "primary" education, a diploma from a pre-*gymnasium* school (in Greece, the *Demotikon*). Information

2.	Total population 10 years of age and over, by sex and general level of education and the proportion illiterate.	19
3.	Graduates of Higher Schools, by sex and type of diploma.	23
4.	Graduates of Higher Schools, by sex and type of (main) diploma. Graduates with a second diploma, by sex and type of this diploma.	24
5.	Persons with diploma or certificate of secondary or elementary technical or other vocational training, by sex and type of diploma.	26
6.	Economically active, by sex, divisions of economic activity[2] and level of general education.	27
7.	Economically active, by groups of individual occupations and general level of education and the share of those who are holders of diploma of technical or other vocational training (secondary or primary).	29
8.	Economically active with diploma (or certificate) of secondary or elementary technical or other vocational education, by sex and divisions of economic activity[2] and the share of those who are unemployed.	31

VOLUME III. EMPLOYMENT (ECONOMICALLY ACTIVE POPULATION, EMPLOYED AND UNEMPLOYED, HOURS OF WORK).

TABLES

Number	Title	Page
1.	Population 10 years of age and over (excluding soldiers and prisoners), by sex and age, classified as economically active (employed and unemployed) and economically inactive.	13
2.	Population 10 years of age and over, by sex, classified as economically active (employed and unemployed) and economically inactive.	17

---

on literacy was obtained by asking respondents: "Do you know how to read and write?" Those with higher, secondary, and elementary education were asked what technical or vocational diplomas they had and from what schools they had received them. Information was also obtained on occupational specialties.

[2]Industrial division.

GREECE

3. Economically active, by sex, age, and divisions of economic activity.[1] ... 21
4. Economically active, by sex, age, and major groups of individual occupations, and the proportion reported as holders of agricultural land or herds. ... 22
5. Economically active, by sex, age, and occupational status. ... 23
6. Economically active, by sex, divisions of economic activity,[1] and major groups of individual occupations. ... 24
7. Economically active, by sex, divisions of economic activity,[1] and occupational status. ... 26
8. Economically active, by sex, major groups of individual occupations and occupational status. ... 27
9. Economically active population (and the proportion who are unemployed), by sex and groups of individual occupations. ... 28
10. Economically active (and the proportion who are unemployed), by sex and groups of individual occupations. ... 30
11. Economically active (and the proportion who are unemployed) by sex and divisions of economic activity.[1] ... 33
12. Economically active (and the proportion who are unemployed) by sex and major groups of individual occupations. ... 35
13. Economically active, by sex and divisions of economic activity.[1] ... 38
14. Economically active, by sex and compound major groups[2] of individual occupations, and the proportion reported as holders of agricultural land or herds. ... 42
15. Economically active, by sex and occupational status. ... 50
16. Employed, by sex, major groups of individual occupations and hours of work during the week prior to census day. Similarly, by marital status for all occupations, as well as for the major group of farmers a.r.w.[3] and the major group of the remaining occupations. ... 54

---

[1] Industrial division.

[2] An abbreviated list of six groups of occupations.

[3] "And related workers," i.e., lumbermen and fishermen.

TABLE TITLES BY VOLUME

VOLUME IV. ECONOMICALLY INACTIVE POPULATION.

TABLES

Number	Title	Page
1.	Economically inactive, by sex, age, and reason for not working.	11
2.	Economically inactive, by sex, age, and main source of sustenance.	15
3.	Economically inactive, by sex, reason for not working, and main source of sustenance.	19
4.	Economically inactive, by sex and reason for not working.	22
5.	Economically inactive, by sex and main source of sustenance.	26

VOLUME V. INTERNAL MIGRATION.[1]

TABLES

Number	Title	Page
1.	Permanent members of private households born before 1956; by sex, by residence in 1961 (urban, semi-urban, or rural) and by residence in 1955; in the same municipality or commune, elsewhere in Greece, or abroad.	39
2.	Permanent members of private households born before 1956 who resided elsewhere in Greece in 1955, by sex and residence (urban, semi-urban, or rural) in 1961 and 1955.	40
3a.	Migration between urban, semi-urban, and rural areas 1956-1960. In-migrants and out-migrants 1956-1960. Net population increase or decrease resulting therefrom.	41
3b.	Migration between urban-rural areas except the migration to, or from, Greater Athens. In-migrants and out-migrants by urban-rural areas. Net increase or decrease which would occur without the simultaneous migration to, or from, Greater Athens.	42

---

[1] Information on internal migration was obtained by asking "In what town or village did you reside at the end of the year 1955?" of all persons born before 1956 and enumerated in private households.

4. Permanent members of private households, born before 1956 and residing elsewhere in Greece in 1955, by sex and geographic region where they resided in 1961 and 1955. 43

5. Migration between geographic regions. In-migrants and out-migrants 1956-1960, by geographic regions. Population increase or decrease resulting therefrom. 44

6. Permanent members of private households born before 1956 and residing in 1955 in a municipality or commune other than that in which they were enumerated in 1961, but within the same geographic region. Distribution by their 1955 and 1961 residence area (urban, semi-urban, or rural). 45

7. Migration between urban-rural areas within each geographic region. In-migrants by, and out-migrants from, each area class. Net population increase or decrease resulting therefrom. 46

8. Permanent members of private households born before 1956 and residing elsewhere in Greece in 1955, by sex and according to topographic characteristics of their residence commune in 1961 and 1955. 47

9. Migration between level, semi-mountainous, and mountainous communes.[2] In-migrants and out-migrants 1955-1960. Net population increase or decrease due to interior migration. 48

10. Permanent members of private households born before 1956 and residing in 1955 in a municipality or commune other than that in which they were enumerated in 1961, but within the same geographic region. Distribution by commune class (level, semi-mountainous, and mountainous) of their 1961 and 1955 residence. 49

11. Migration between level, semi-mountainous, and mountainous communes within each geographic region. In-migrants to, and out-migrants from, each commune class. Net population increase or decrease resulting therefrom. 50

---

[2]Level communes are less than 800 meters above sea level, and their territory exhibits altitude differences not larger than 100 or 150 meters. The greater part of the territory of semi-mountainous communes is less than 800 meters above sea level and exhibits altitude differences of 300 to 400 meters. Mountainous communes are more than 800 meters above sea level and exhibit altitude differences of more than 400 meters.

TABLE TITLES BY VOLUME

12. In-migrants to, and out-migrants from, communes belonging to urban, semi-urban, and rural areas, by sex and age.   51

13. Total population 10 years of age and over (except soldiers and prisoners), and permanent members of private households, 10 years of age and over, resident in 1955: (a) in the same commune, (b) elsewhere in Greece. Economically active among these, and, among the economically active, number of unemployed, by sex and urban-rural areas.   52

14. Economically active permanent members of private households and unemployed among these, by sex, residence in 1955 (in the same commune or elsewhere in Greece), and branch of economic activity.   53

IRELAND (EIRE)

MAIN ENTRY AND VOLUME NOTES:

Ireland (Eire) Central Statistics Office.
   Census of population of Ireland, 1961.   Dublin, 1963-
     v. 31cm.   (66-56622)

Vol. I. Population, area, and valuation of each district electoral division and of each larger unit of area. 1963. xi, 152 p.

Vol. II. Ages and conjugal conditions classified by areas only. 1963. vi, 221 p.

Vol. III. Occupations of males and females in each province, county, county borough, and in each town of 5,000 and over population. 1963. vi, 172 p.

Vol. IV. Industries. 1964. vii, 147 p.

Vol. V. Occupations classified by ages and conjugal conditions. 1964. vi, 205 p.

Vol. VI. Housing and social amenities. 1964. vii, 165 p. [Excluded.]

Vol. VII. Part I. Religions. Part II. Birthplaces. 1965. iv, 117 p.

Vol. VIII. Fertility of marriage. 1965. iv, 215 p.

Vol. IX. Irish language with special tables for the Gaeltacht areas. 1966. v, 35 p.

ENUMERATION DATE: April 9, 1960.

ENUMERATION BASIS: De facto.

GLOSSARY

ADMINISTRATIVE, GEOGRAPHICAL, AND POLITICAL UNITS

Level	Vernacular Term
National	Republic of Ireland
Intermediate	Province
Local	County/County borough

GLOSSARY

                    Urban/(former) Rural district[1]
                    District electoral division

Public health concepts:

                    County/County borough
                    Superintendent registrar's district
                    Registrar's district

Electoral divisions:

                    Constituency
                    County electoral area
                    District/Borough electoral area
                    Ward

Ecclesiastical:

                    Civil parish

Other concepts:

                    Town
                    Census town
                    City
                    Type of district

    Ireland (Eire) is divided into four provinces (Leinster, Munster, Connaught, and Ulster), which no longer have any political significance. These are divided into 27 administrative counties and the four county boroughs of Dublin, Cork, Waterford, and Limerick.

    Counties consist of urban and (former) rural districts, which in turn are composed of district electoral divisions. The counties have county councils with administrative responsibility for municipal boroughs,[2] urban districts, towns, and rural districts.

    County boroughs have single councils responsible for all services within their areas and independent of the counties.

    The smallest territorial divisions for which population statistics are published are the district electoral divisions or, in the county boroughs, the wards.

---

[1] Rural districts were abolished as administrative units in 1925, but have been retained for census purposes (see Explanatory Notes, Volume I, p. vi, of the census).

[2] Term not used in table titles.

IRELAND (EIRE)

Public health services are organized on a county and county borough basis. To provide these services, counties and county boroughs are divided into <u>superintendent registrar's districts</u>, which in turn are made up of <u>registrar's districts</u> and dispensary districts.[2]

For Parliamentary elections, the country is divided into <u>constituencies</u>. For elections to county councils, counties are divided into <u>county electoral areas</u>, which are groups of district electoral divisions.

<u>Civil parishes</u> were originally ecclesiastical divisions that were later used for civil purposes. They were deleted as obsolete from censuses between 1911 and 1950. The current census resumes tabulations for the civil parishes of Dublin County and County Borough, and also certain civil parishes in or adjacent to the county boroughs of Cork, Limerick, and Waterford.

<u>Town</u> is a general term. Towns fall into two classes--those with and those without legally defined boundaries. The first class comprises (a) the 4 county boroughs and the borough of Dun Laoghaire, (b) the 6 municipal boroughs, (c) the 49 urban districts, and (d) the 29 towns under the Towns Improvement Act (Ireland, 1854). The areas enumerated as towns without legally defined boundaries are determined by the census authorities and are referred to as <u>census towns</u>.

<u>City</u> is an historical or ceremonial term for certain towns, and has no administrative or political significance.

<u>Type of district</u> refers to a division of the country into the Dublin region, other county boroughs, towns by size-class, and rural districts. The size-classes of towns range in population from 500-1500 to 10,000 and over.

TABLE OF DEMOGRAPHIC CONCEPTS

The demographic concepts which appear in the following table correspond to the recommendations of the European Conference of Statisticians, unless otherwise noted. They are defined and discussed in the appendixes and in footnotes. Terms in brackets are not used in the table titles, but they are included here in order to give logical completeness to the various classifications in the table.

TABLE TITLES BY VOLUME

Appendix	English Table Term	Vernacular Table Term
ECONOMIC CHARACTERISTICS		
A	Economic activity Economically active Economically inactive	Same Gainfully occupied[1] Not gainfully occupied
B	Employment status	Same
C	Industrial classification [Division] Major group(s) Group(s)	Same ... Industrial group Industry
D	Occupational classification Major group(s) Minor group(s)	Same Occupational group Occupation
E	Socio-economic category	Social group[2]
HOUSEHOLD AND FAMILY CHARACTERISTICS		
F	Family	Same

TABLE TITLES BY VOLUME

VOLUME I. POPULATION, AREA, AND VALUATION OF EACH DISTRICT ELECTORAL DIVISION AND OF EACH LARGER UNIT OF AREA.

TABLES

Number	Title	Page
1.	Population of each province at each census since 1901; marriages, births, and deaths registered, natural increase and estimated net emigration in each intercensal period since 1891.	2
2.	Average annual number of marriages, births, deaths, etc., per 1,000 of the average population in each province for each intercensal period since 1901.	3

---

[1] Persons 14 years of age and older who are actively engaged in an occupation from which they obtain their livelihood (employed and unemployed), excluding children seeking their first job, who are counted as economically inactive.

[2] A detailed schedule showing the distribution of occupations in social groups is provided in Appendix C of Volume III of this census.

IRELAND (EIRE)

3.	Population of each province and county as constituted at each of the eleven censuses from 1871 to 1961.	4
4.	Increase or decrease per cent in population of each province and county in each intercensal period from 1871 to 1961.	7
5.	Population of each constituency for elections to Dáil Eireann.	8
6.	Population, area, and valuation of each county electoral area and each borough electoral area.	9
7.	Population, area, and valuation of boroughs (not being county boroughs), urban districts, and other towns possessing local government.	13
8.	Population of cities and towns in each county on 9 April, 1961.	15
9.	Population, area, and valuation of towns with legally defined boundaries; population of environs of towns with legally defined boundaries; population of towns without legally defined boundaries having 1,500 inhabitants or over in 1961; total town population and number of persons per 100 acres.	20
10.	Population, area, and valuation of superintendent registrars' districts and registrars' districts on 9 April, 1961, and the number of persons per 100 acres.	38
11.	Population, area, and valuation of each district electoral division, urban district, rural district, and county on 9 April, 1961.	56
12.	Births, deaths, natural increase, and estimated net emigration in the intercensal period 1956-1961 for each province and county and average annual rates per 1,000 average population.	138
13.	Population of towns by type of district on 9 April, 1961.	140
14A.	Population in 1901 and 1961 of civil parishes in Dublin County and County Borough.	142
14B.	Population in 1901 and 1961 of certain civil parishes in or adjacent to the county boroughs of Cork, Limerick, and Waterford.	144
15.	Alphabetical list of towns, with populations, 1961.	146

TABLE TITLES BY VOLUME

VOLUME II. AGES AND CONJUGAL CONDITIONS CLASSIFIED BY AREAS ONLY.[1]

TABLES

Number	Title	Page
1A.	Number of persons in each age group, classified by conjugal condition, at each census--1926 to 1961.	1
1B.	Number of males in each age group, classified by conjugal condition, at each census--1926 to 1961.	2
1C.	Number of females in each age group, classified by conjugal condition, at each census--1926 to 1961.	3
2.	Single males and females as a percentage of the total in each age group at each census--1926 to 1961.	4
3.	Percentage change in number of persons, males and females, in each age group in the intercensal periods 1936-1946, 1946-1951, and 1951-1961.	5
4.	Average annual rate of change in number of persons, males and females, per 1,000 average population in each age group, in the intercensal periods 1936-1946, 1946-1951, and 1951-1961.	5
5A.	Number of males in each age group, per 10,000 males in each county and county borough.	6
5B.	Number of females in each age group, per 10,000 females in each county and county borough.	7
6A.	Single males in each age group, as a percentage of the total in that age group, in each county and county borough.	8
6B.	Single males in each age group, as a percentage of the total in that age group, in the aggregate town areas in each county and county borough.	9
6C.	Single males in each age group, as a percentage of the total in that age group, in the aggregate rural areas in each county.	10
7A.	Single females in each age group, as a percentage of the total in that age group, in each county and county borough.	11
7B.	Single females in each age group, as a percentage of the total in that age group, in the aggregate town areas in each county and county borough.	12

---

[1] Age data were obtained by asking the date of birth--day, month, and year--rather than the age of the person or the number of complete years lived. The small number of persons enumerated as "separated" or "divorced" were classified as "married."

IRELAND (EIRE)

7C.	Single females in each age group, as a percentage of the total in that age group, in the aggregate rural areas in each county.	13
8A.	Females, per 1,000 males in each age group, in each county and county borough.	14
8B.	Females, per 1,000 males in each age group, in the aggregate town areas, in each county and county borough.	15
8C.	Females, per 1,000 males in each age group, in the aggregate rural areas in each county.	16
9.	Population classified by sex, year of age, and conjugal condition.	17
10.	Total population at, or over, each year of age.	19
11.	Males and females in each province, county, and county borough, classified by year of age--0 to 69 years.	20
12.	Population of each province, county, and county borough, classified by sex, age group, and conjugal condition.	29
13.	Males and females classified by age group and conjugal condition in urban districts, other towns with legally defined boundaries, environs of towns with legally defined boundaries, towns without legally defined boundaries having 1,500 inhabitants or over, and rural districts.	48
14.	Males and females in each ward in Dublin County Borough, classified by age group and conjugal condition.	144
15.	Males and females in each superintendent registrar's district and each registrar's district, classified by age group.	155
16.	Population of aggregate town areas and aggregate rural areas, classified by sex, age group, and conjugal condition.	221

VOLUME III. OCCUPATIONS OF MALES AND FEMALES IN EACH PROVINCE, COUNTY, COUNTY BOROUGH, AND IN EACH TOWN OF 5,000 AND OVER POPULATION.

TABLES

Number	Title	Page
1.	Numbers of persons, males and females, in each occupational group and numbers in each group per 1,000 gainfully occupied.	2

TABLE TITLES BY VOLUME

2.	Persons, males and females, classified by occupation.	3
3.	Persons, males and females, classified by occupation in 1951 and 1961.	11
4A.	Persons in each province, county, and county borough classified by occupational group.	16
4B.	Males in each province, county, and county borough classified by occupational group.	18
4C.	Females in each province, county, and county borough classified by occupational group.	20
5.	Males and females in each province, county, and county borough classified by occupation.	22
6A.	Males in the principal occupations in each city or town (including suburbs or environs, if any) having a population of 5,000 or over.	108
6B.	Females in the principal occupations in each city or town (including suburbs or environs, if any) having a population of 5,000 or over.	116
7.	Population of each province, county, and county borough, classified by sex and social group, distinguishing gainfully occupied, not gainfully occupied, and children under 14 years of age.	120
8.	Males and females gainfully occupied and not gainfully occupied and children under 14 years of age in each province, county, and county borough in the social group "farmers, farmers' relatives, and farm managers," classified by rateable valuation.[1]	126
9A.	Males in occupational groups classified by rateable valuation.[1]	132
9B.	Females in occupational groups classified by rateable valuation.[1]	142
10.	Male farmers and farmers' relatives having a subsidiary occupation, classified by size of farm and by subsidiary occupation.	152
11.	Landholders, other than farmers, classified by size of holding and occupation.	157

---

[1] Each household was asked the area and "rateable valuation" of all agricultural holdings (if any) of which persons usually resident in the household were the rated occupants. From this, a further subdivision of the social group--"farmers, relatives assisting, and farm managers"--was made according to the rateable valuation of all agricultural holdings owned.

IRELAND (EIRE)

12A.	Males classified by occupation and employment status.	162
12B.	Females classified by occupation and employment status.	167

VOLUME IV. INDUSTRIES.

TABLES

Number	Title	Page
1.	Persons, males and females at work,[1] classified by industrial group, with numbers in each group per 1,000 in all industries.	1
2.	Persons, males and females at work,[1] classified by industry.	2
3.	Persons, males and females, classified by industry in 1951 and 1961.	7
4.	Males and females at work[1] in each province, county, and county borough, classified by industrial group.	12
5.	Males and females at work[1] in each province, county, and county borough, classified by industry.	18
6.	Males and females at work[1] in each industry, distinguishing the principal occupations within that industry.	72
7.	Employment status of persons, males and females, in 1961, with comparative figures for 1951.	128
8A.	Persons classified by industrial group and by employment status.	129
8B.	Males classified by industrial group and by employment status.	130
8C.	Females classified by industrial group and by employment status.	131
9.	Males and females classified by industry and by employment status.	132
10.	Employment status of males and of females in each province, county, and county borough.	144

---

[1] "At work" means employed.

TABLE TITLES BY VOLUME

VOLUME V. OCCUPATIONS CLASSIFIED BY AGES AND CONJUGAL CONDITIONS.

TABLES

Number	Title	Page
1A.	Males in each occupational group classified by age group and conjugal condition.	1
1B.	Females in each occupational group classified by age group and conjugal condition.	4
2A.	Males in each occupation classified by age group and conjugal condition.	6
2B.	Females in each occupation classified by age group and conjugal condition.	32
3A.	Males in each occupational group in each province, county, and county borough, classified by age group and conjugal condition.	45
3B.	Females in each occupational group in each province, county, and county borough, classified by age group and conjugal condition.	82
4.	Farmers in each province and county classified by size of farm, age group, and conjugal condition.	119
5A.	Males classified by occupation and by individual year of age from 14 to 19.	133
5B.	Females classified by occupation and by individual year of age from 14 to 19.	137
6A.	Males, aged 14 to 19 years, in each major occupational group in each county and county borough, classified by individual ages.	140
6B.	Females, aged 14 to 19 years, in each major occupational group in each county and county borough, classified by individual ages.	156
7.	Total population in each social group, classified by age group and conjugal condition, distinguishing sex, gainfully occupied, and not gainfully occupied for those aged 14 years and over.	172
8.	Total population in the social group "farmers, farmers' relatives, and farm managers," classified by rateable valuation[1] of agricultural holding, and by age group and conjugal condition, distinguishing sex, gainfully occupied, and not gainfully occupied for those aged 14 years and over.	178

---

[1] See footnote 1, p. 207.

IRELAND (EIRE)

9A.	Males classified by employment status and by age group.	184
9B.	Males in each age group as a percentage of the total in each employment status.	184
9C.	Males classified by employment status in each age group expressed as a percentage of the total gainfully occupied in that age group.	184
10A.	Females classified by employment status and by age group.	185
10B.	Females in each age group as a percentage of the total in each employment status.	185
10C.	Females classified by employment status in each age group expressed as a percentage of the total gainfully occupied in that age group.	185
11.	Males and females aged 14 to 19 years classified by employment status.	186
12A.	Males in each employment status classified by age group and conjugal condition.	187
12B.	Females in each employment status classified by age group and conjugal condition.	188
13A.	Males at work in each industrial group classified by age group.	189
13B.	Females at work in each industrial group classified by age group.	190
14A.	Males at work in each industry classified by age group.	191
14B.	Females at work in each industry classified by age group.	199

VOLUME VII. PART I. RELIGIONS. PART II. BIRTHPLACES.

TABLES

Number	Title	Page
Part I. Religions[1]		
1A.	Persons of each religion at each census--1881 to 1961.	1

---

[1] Classifications of the following religious denominations are given: Catholic, Church of Ireland, Presbyterian, Methodist, Jewish, and Baptist.

TABLE TITLES BY VOLUME

1B.	Percentage increase or decrease in number of persons of each religion in intercensal periods, 1881 to 1961.	1
1C.	Persons of each religion as a percentage of the total population--1881 to 1961.	1
2A.	Males and females of each religion at each census--1881 to 1961.	2
2B.	Percentage increase or decrease in number of males and females of each religion in intercensal periods, 1881 to 1961.	2
2C.	Males and females of each religion as a percentage of the total population--1881 to 1961.	2
3.	Persons of each religion in each province and county at each census--1881 to 1961.	4
4A.	Persons, males and females, of each religion in each province.	9
4B.	Percentage increase or decrease in number of persons, males and females, of each religion in each province--1946 to 1961.	9
4C.	Persons, males and females, of each religion as a percentage of the total population in each province.	9
5A.	Persons of each religion in the aggregate town areas and aggregate rural areas of each province.	10
5B.	Persons of each religion as a percentage of the total population in the aggregate town areas and aggregate rural areas of each province.	10
6A.	Persons of each religion in each type of district.	11
6B.	Persons of each religion as a percentage of the total population in each type of district.	11
7A.	Persons of each religion in each province, county, and county borough.	12
7B.	Percentage increase or decrease in number of persons of each religion in each province, county, and county borough--1946 to 1961.	13
7C.	Persons of each religion as a percentage of the total population in each province, county, and county borough.	14
8.	Males and females of each religion in urban districts, other towns with legally defined boundaries, environs of towns with legally defined boundaries, towns without legally defined boundaries having 1,500 inhabitants or over, and rural districts.	15

IRELAND (EIRE)

9A.	Persons of each religion classified by age group and conjugal condition.	36
9B.	Males of each religion classified by age group and conjugal condition.	37
9C.	Females of each religion classified by age group and conjugal condition.	38
10.	Males and females in each province, county, and county borough, classified by religion, age group, and conjugal condition.	39
11A.	Males in occupational groups classified by religion.	75
11B.	Females in occupational groups classified by religion.	77

Part II. Birthplaces[2]

1A.	Persons, males and females, in each province on 9 April, 1961, classified by country of birth.	78
1B.	Persons, males and females, in the aggregate town areas of each province on 9 April, 1961, classified by country of birth.	79
1C.	Persons, males and females, in the aggregate rural areas of each province on 9 April, 1961, classified by country of birth.	80
2A.	Persons in each province, county, and county borough, classified by country of birth.	81
2B.	Males in each province, county, and county borough, classified by country of birth.	82
2C.	Females in each province, county, and county borough, classified by country of birth.	83
3.	Persons in each province and county classified by country of birth--1926 to 1961.	84
4.	Males and females in each province, county, and county borough, classified by county and country of birth.	86
5.	Males and females born in each county classified by place of residence.	95
6.	Persons in each city or town (including suburbs or environs, if any) of 1,500 or more inhabitants, classified by country of birth.	96

---

[2] Respondents who were born in a hospital were instructed to give the county of residence of their mother as their place of birth.

7.	Persons, males and females, classified by country of birth and age group.	98
8.	Males and females in each province, county, and county borough, classified by country of birth and age group.	99
9A.	Males in occupational groups, classified by country of birth.	111
9B.	Females in occupational groups, classified by country of birth.	113
10.	Males and females born outside All Ireland, classified by country of birth and year of taking up residence.	114
11.	Persons born outside All Ireland, resident in each province, county, and county borough, classified by year of taking up residence.	116

VOLUME VIII. FERTILITY OF MARRIAGE.[1]

TABLES

Number	Title	Page
1.	Families and children born--provinces.	2
1A.	Distribution of families and of children born according to number of children in family--provinces.	3
2.	Families classified by duration of marriage and by number of children born.	4
2A.	Distribution of families in each duration of marriage group according to number of children in family.	4
3.	Children born, classified by duration of marriage and by number of children born.	5

---

[1] The inquiry on fertility covered all married women in the country whether or not their husbands were residing with them at the date of the census. The information collected was on existing marriages only (widows and single women were excluded), and the question on the number of children born alive excluded children of previous marriages, stepchildren, and illegitimate children. Information on the date of present marriage and the date of birth of each married woman made it possible to calculate the <u>duration of marriage</u> in completed years and the <u>age at marriage</u> of the wife. Similarly, when the husband was enumerated on the same schedule as his wife, his date of birth made possible a calculation of his age at marriage.

IRELAND (EIRE)

3A. Distribution of children born in each duration of marriage group, according to number of children in family.   5

4. Families and children born, classified by duration of marriage and age of wife at marriage.   6

4A. Average number of children born per 100 families in each duration of marriage, by age of wife at marriage group.   6

5. Families and children born, classified by age of wife at marriage and age of husband at marriage.   7

5A. Average number of children born per 100 families in each age of wife at marriage, by age of husband at marriage group.   7

6. Families, children born, and average number of children born per 100 families in each duration of marriage group--Catholic and other religious denominations.   8

7. Families classified by duration of marriage, age of wife at marriage, age of husband at marriage, and number of children born.   9

7A. Average number of children born per 100 families in each duration of marriage, by age of wife at marriage, by age of husband at marriage group.   34

7B. Percentage of childless marriages in each duration of marriage, by age of wife at marriage, by age of husband at marriage group--duration ten years and over.   37

8. Families classified by duration of marriage, age of wife at marriage, and number of children born--provinces.   39

8A. Average number of children born per 100 families in each duration of marriage, by age of wife at marriage group--provinces.   63

9. Families classified by duration of marriage, age of wife at marriage, and number of children born--type of district.   67

9A. Average number of children born per 100 families in each duration of marriage, by age of wife at marriage group--type of district.   91

10. Families and children born, classified by duration of marriage and age of wife at marriage--provinces, counties, and county boroughs.   94

10A. Average number of children born per 100 families in each duration of marriage, by age of wife at marriage group--provinces, counties, and county boroughs.   109

11.	Families classified by duration of marriage, age of wife at marriage, and number of children born--Catholic and other religious denominations.	124
11A.	Average number of children born per 100 families in each duration of marriage, by age of wife at marriage group--Catholic and other religious denominations.	131
11B.	Percentage of childless marriages in each duration of marriage, by age of wife at marriage group--Catholic and other religious denominations for durations ten years and over.	132
12.	Families and children born, classified by duration of marriage and age of wife at marriage--provinces and religious denominations.	133
12A.	Average number of children born per 100 families in each duration of marriage, by age of wife at marriage group--provinces and religious denominations.	141
13.	Families and children born, classified by duration of marriage and age of wife at marriage--religious denominations.	145
13A.	Average number of children born per 100 families in each duration of marriage, by age of wife at marriage group--religious denominations.	149
14.	Families classified by duration of marriage, age of wife at marriage, and number of children born--social groups.	152
14A.	Average number of children born per 100 families in each duration of marriage, by age of wife at marriage group--social groups.	184
14B.	Percentage of childless marriages in each duration of marriage, by age of wife at marriage group--social groups for durations ten years and over.	188
15.	Families and children born, classified by duration of marriage and age of wife at marriage--social groups and religious denominations.	190
15A.	Average number of children born per 100 families in each duration of marriage, by age of wife at marriage group--social groups and religious denominations.	201
16.	Families classified by duration of marriage, present age of wife, and number of children born.	207
17.	Families classified by present age of wife and number of children born.	212
18.	Families classified by year of marriage and number of children born.	214

IRELAND (EIRE)

VOLUME IX. IRISH LANGUAGE WITH SPECIAL TABLES FOR THE GAELTACHT AREAS.[1]

TABLES

Number	Title	Page
1.	Irish speakers and non-Irish speakers in each province, at each census from 1861 to 1946 and 1961.	2
2.	Irish speakers as percentage of total persons in each province, at each census from 1861 to 1946 and 1961.	2
3.	Number and percentage of Irish speakers, 3 years of age and over, in each province and county--1936, 1946, and 1961.	3
4.	Number and percentage of Irish speakers in each urban and rural district in 1961, with comparative percentage figures for 1946--persons 3 years of age and over.	4
5.	Number and percentage of Irish speakers and non-Irish speakers, 3 years of age and over, in each province, classified by age group--1936, 1946, and 1961.	13
6.	Irish speakers and non-Irish speakers, 3 years of age and over, in each province, county, and county borough, classified by age group.	14
6A.	Irish speakers as a percentage of the total in each age group.	18
7.	Irish speakers classified by occupational group.	20
8.	Total population (all ages) in Fior-Ghaeltacht and Breac-Ghaeltacht areas, as defined in the Gaeltacht Commission Report, 1925, at each census--1926 to 1961.	24

---

[1] This volume classifies the population 3 years of age and over by knowledge of the Irish language. In answer to the question, "Ability to speak the Irish language?," persons were instructed to write "Irish only" if they could speak only Irish, "Irish and English" if they could speak Irish and English, "Read but cannot speak Irish" if they could read but not speak Irish. Persons who could not read or speak Irish were instructed to leave the question blank. The Gaeltacht areas, as designated in the Gaeltacht Areas Order, 1956, are Fior-Ghaeltacht and Breac-Ghaeltacht. Fior-Ghaeltacht contains portions of the counties of Clare, Cork, Donegal, Galway, Kerry, Mayo, and Waterford. Breac-Ghaeltacht includes portions of the foregoing counties, and also small areas in the counties of Sligo and Tipperary.

9.	Irish speakers and non-Irish speakers, 3 years of age and over, in the Fíor-Ghaeltacht and Breac-Ghaeltacht areas, as defined in the Gaeltacht Commission Report, 1925, in each county--1936, 1946, and 1961.	25
10.	Total population (all ages) in 1956 and 1961 of the portion of each county designated as a Gaeltacht area in the Gaeltacht Areas Order, 1956.	26
11.	Irish speakers and non-Irish speakers, 3 years of age and over, in 1961, in the portion of each county designated as a Gaeltacht area in the Gaeltacht Areas Order, 1956.	26
12A.	Irish speakers and non-Irish speakers, 3 years of age and over, classified by age group, in the portion of each county designated as a Gaeltacht area in the Gaeltacht Areas Order, 1956.	27
12B.	Males, Irish speakers and non-Irish speakers, 3 years of age and over, classified by age group, in the portion of each county designated as a Gaeltacht area in the Gaeltacht Areas Order, 1956.	28
12C.	Females, Irish speakers and non-Irish speakers, 3 years of age and over, classified by age group, in the portion of each county designated as a Gaeltacht area in the Gaeltacht Areas Order, 1956.	29
13.	Total population in 1956 and 1961 and total persons 3 years of age and over in 1961, distinguishing Irish speakers and non-Irish speakers, in district electoral divisions or parts of district electoral divisions within the Gaeltacht as defined by the Gaeltacht Areas Order, 1956.	30

ITALY

MAIN ENTRIES AND VOLUME NOTES:

Italy. Istituto Centrale di Statistica.
10° [i.e., Decimo] censimento generale della popolazione,
15 ottobre 1961. Roma, 1963-
v. 32cm. (NUC64-6925)

Vol. I. Dati riassuntivi comunali e provinciali sulla popolazione e sulle abitazioni. 1963. 156 p.

Vol. II. Dati riassuntivi comunali e provinciali su alcune principali caratteristiche strutturali della popolazione, sesso, età, istruzione, attività economica. 1963. 533 p.

Vol. III. Dati sommari per comune. [Fascicoli 1-92]

Fasc. 1. Provincia di Torino. 1966. 171 p.
Fasc. 2. Provincia di Vercelli. 1965. 103 p.
Fasc. 3. Provincia di Novara. 1966. 103 p.
Fasc. 4. Provincia di Cuneo. 1966. 143 p.
Fasc. 5. Provincia di Asti. 1964. 73 p.
Fasc. 6. Provincia di Alessandria. 1966. 105 p.
Fasc. 7. Valle d'Aosta. 1964. 45 p.
Fasc. 8. Provincia di Imperia. 1964. 43 p.
Fasc. 9. Provincia di Savona. 1964. 45 p.
Fasc. 10. Provincia di Genova. 1966. 49 p.
Fasc. 11. Provincia di La Spezia. 1964. 43 p.
Fasc. 12. Provincia di Varese. 1966. 71 p.
Fasc. 13. Provincia di Como. 1966. 131 p.
Fasc. 14. Provincia di Sondrio. 1966. 69 p.
Fasc. 15. Provincia di Milano. 1966. 137 p.
Fasc. 16. Provincia di Bergamo. 1966. 133 p.
Fasc. 17. Provincia di Brescia. 1966. 133 p.
Fasc. 18. Provincia di Pavia. 1966. 109 p.
Fasc. 19. Provincia di Cremona. 1966. 75 p.
Fasc. 20. Provincia di Mantova. 1966. 51 p.
Fasc. 21. Provincia di Bolzano, Provinz Bozen. 1964. 229 p.
Fasc. 22. Provincia di Trento. 1964. 133 p.
Fasc. 23. Provincia di Verona. 1965. 81 p.
Fasc. 24. Provincia di Vicenza. 1965. 83 p.
Fasc. 25. Provincia di Belluno. 1964. 51 p.
Fasc. 26. Provincia di Treviso. 1965. 75 p.
Fasc. 27. Provincia di Venezia. 1965. 45 p.
Fasc. 28. Provincia di Padova. 1965. 75 p.

Fasc. 29. Provincia di Rovigo. 1964. 45 p.
Fasc. 30. Provincia di Udine. 1965. 119 p.
Fasc. 31. Provincia di Gorizia. 1964. 27 p.
Fasc. 32. Provincia di Trieste. 1965. 31 p.
Fasc. 33. Provincia di Piacenza. 1964. 53 p.
Fasc. 34. Provincia di Parma. 1964. 57 p.
Fasc. 35. Provincia di Reggio nell'Emilia. 1964. 49 p.
Fasc. 36. Provincia di Modena. 1964. 53 p.
Fasc. 37. Provincia di Bologna. 1965. 53 p.
Fasc. 38. Provincia di Ferrara. 1964. 35 p.
Fasc. 39. Provincia di Ravenna. 1965. 29 p.
Fasc. 40. Provincia di Forlì. 1965. 45 p.
Fasc. 41. Provincia di Pesaro e Urbino. 1964. 49 p.
Fasc. 42. Provincia di Ancona. 1964. 47 p.
Fasc. 43. Provincia di Macerata. 1964. 47 p.
Fasc. 44. Provincia di Ascoli Piceno. 1965. 45 p.
Fasc. 45. Provincia di Massa-Carrara. 1964. 33 p.
Fasc. 46. Provincia di Lucca. 1964. 53 p.
Fasc. 47. Provincia di Pistoia. 1964. 33 p.
Fasc. 48. Provincia di Firenze. 1965. 53 p.
Fasc. 49. Provincia di Livorno. 1964. 27 p.
Fasc. 50. Provincia di Pisa. 1964. 45 p.
Fasc. 51. Provincia di Arezzo. 1965. 51 p.
Fasc. 52. Provincia di Siena. 1964. 45 p.
Fasc. 53. Provincia di Grosseto. 1964. 33 p.
Fasc. 54. Provincia di Perugia. 1965. 59 p.
Fasc. 55. Provincia di Terni. 1965. 43 p.
Fasc. 56. Provincia di Viterbo. 1965. 41 p.
Fasc. 57. Provincia di Rieti. 1965. 47 p.
Fasc. 58. Provincia di Roma. 1966. 71 p.
Fasc. 59. Provincia di Latina. 1965. 39 p.
Fasc. 60. Provincia di Frosinone. 1965. 71 p.
Fasc. 61. Provincia di Caserta. 1965. 69 p.
Fasc. 62. Provincia di Benevento. 1966. 67 p.
Fasc. 63. Provincia di Napoli. 1966. 69 p.
Fasc. 64. Provincia di Avellino. 1966. 69 p.
Fasc. 65. Provincia di Salerno. 1965. 101 p.
Fasc. 66. Provincia di L'Aquila. 1965. 73 p.
Fasc. 67. Provincia di Teramo. 1966. 45 p.
Fasc. 68. Provincia di Pescara. 1965. 41 p.
Fasc. 69. Provincia di Chieti. 1965. 69 p.
Fasc. 70. Provincia di Campobasso. 1965. 69 p.
Fasc. 71. Provincia di Foggia. 1965. 41 p.
Fasc. 72. Provincia di Bari. 1966. 39 p.
Fasc. 73. Provincia di Taranto. 1966. 27 p.
Fasc. 74. Provincia di Brindisi. 1965. 25 p.
Fasc. 75. Provincia di Lecce. 1966. 65 p.
Fasc. 76. Provincia di Potenza. 1965. 67 p.
Fasc. 77. Provincia di Matera. 1964. 37 p.
Fasc. 78. Provincia di Cosenza. 1966. 99 p.
Fasc. 79. Provincia di Catanzaro. 1966. 97 p.

Fasc. 80. Provincia di Reggio di Calabria. 1965. 71 p.
Fasc. 81. Provincia di Trapani. 1966. 31 p.
Fasc. 82. Provincia di Palermo. 1966. 67 p.
Fasc. 83. Provincia di Messina. 1966. 73 p.
Fasc. 84. Provincia di Agrigento. 1965. 39 p.
Fasc. 85. Provincia di Caltanissetta. 1965. 27 p.
Fasc. 86. Provincia di Enna. 1965. 25 p.
Fasc. 87. Provincia di Catania. 1965. 41 p.
Fasc. 88. Provincia di Ragusa. 1965. 25 p.
Fasc. 89. Provincia di Siracusa. 1965. 27 p.
Fasc. 90. Provincia di Sassari. 1965. 65 p.
Fasc. 91. Provincia di Nuoro. 1966. 65 p.
Fasc. 92. Provincia di Cagliari. 1966. 95 p.

Vol. III. Dati sommari per comune. Appendice. Dati riassuntivi nazionali. 1966. 59 p.

Vol. IV. Famiglie e convivenze. 1967. 697 p.

Vol. V. Sesso, età, stato civile, luogo di nascita. 1968. 733 p.

Vol. VI. Professioni. 1967. 1204 p.

Vol. VII. Istruzione. 1968. 722 p.

Vol. VIII. Abitazioni. 1967. 907 p. [Excluded.]

Vol. IX. Dati generali riassuntivi. 1969. 584 p.

Italy. Istituto Centrale di Statistica.
10° [i.e., Decimo] censimento generale della popolazione, 15 ottobre 1961; popolazione legale dei comuni. Roma, 1963. 241 p. 21cm.

ENGLISH TRANSLATION OF CENSUS AND VOLUME TITLES:

Tenth general census of the population, 15 October 1961.

I. Summary data by commune and province on population and housing.

II. Summary data by commune and province on some principal structural characteristics of the population: sex, age, level of education, and economic activity.

III. Summary data by commune. [92 fascicles]
   Appendix: Summarized national data.

IV. Households and institutions.

V. Sex, age, marital status, place of birth.

VI. Occupations.

VII. Education.

GLOSSARY

VIII. Housing.

IX. General summary data.

Tenth general census of the population, 15 October 1961; resident population of the communes.

ENUMERATION DATE: October 15, 1961.

ENUMERATION BASIS: De facto; de jure.

GLOSSARY

ADMINISTRATIVE, GEOGRAPHICAL, AND POLITICAL UNITS

Level	English Translation	Vernacular Term
National	Nation (national)	Stato (nazionale)
Intermediate	Region (regional)	Regione (regionale)
	Province (provincial)	Provincia (provinciale)
Local	Commune (communal)	Comune (comunale)
	Urban	Urbano
	Rural	Rurale
Other concepts:		
	Altimetric zone	Zona altimetrica
	Commune size-class	Comune per classe di ampiezza demographica
	Density	Densità
	Geographical sections	Frazioni geografiche

Italy is divided into 20 <u>regions</u>. The regions correspond to areas of the country with specific population, climatic, and industrial characteristics.

The regions are divided into a total of 92 <u>provinces</u>.

Provinces are divided into <u>communes</u>, the basic units of local government. Communes vary in size and population, from rural settlements to large cities serving as provincial capitals.

<u>Altimetric zones</u> are areas of mountains, hills, and plains within each province and region.

ITALY

Communes are grouped into <u>commune size-classes,</u> which range in population from under 500 to over 500,000.

<u>Density</u> refers to the population per square kilometer.

<u>Geographical sections</u> are parts of communal territory normally consisting of a population center, with population nuclei and scattered houses around that center.

TABLE OF DEMOGRAPHIC CONCEPTS

The demographic concepts which appear in the following table correspond to the recommendations of the European Conference of Statisticians, unless otherwise noted. They are defined and discussed in the appendixes and in footnotes. Terms in parentheses are alternative forms which appear in the table titles. Terms in brackets are not used in the table titles, but they are included here in order to give logical completeness to the various classifications in the table.

Appendix	English Table Term	Vernacular Table Term
\multicolumn{3}{l}{CULTURAL, EDUCATIONAL, AND PERSONAL CHARACTERISTICS}		
...	Age[1]	Età
	Level of education[2]	Grado d'istruzione
	University graduates	Laureati
	Diplomates	Diplomati
	Persons with certificate from secondary school	Licenza di scuola media inferiore
	Persons with certificate from elementary school	Titolo di scuola elementare
	Persons lacking certification	Privi di titolo studio
\multicolumn{3}{l}{ECONOMIC CHARACTERISTICS}		
A	[Economic activity]	...
	Economically active	Popolazione attiva
	Economically inactive	Popolazione non attiva
B	Employment status	Posizione nella professione
C	Industrial classification[3]	Attività economica
	Division(s)	Ramo
	Major group(s)	Classe
	Group(s)	Sottoclasse
	Subgroup(s)	Categoria

---

[1][All footnotes are at the end of the Table.]

GLOSSARY

D	Occupational classification[4]	Professione
	Major group (Occupation)	Professione
	Minor group (Occupation)	Professione
	Unit group (Occupation)	Professione
	Occupational/Non-occupational status[5]	Condizione professionale o non professionale
ENUMERATION BASIS		
	Enumerated population[6]	Popolazione presente
	Resident population[7]	Popolazione residente
HOUSEHOLD AND FAMILY CHARACTERISTICS		
F	[Household]	...
	Private household[8]	Famiglia
	Institutional household[9]	Convivenza
	Head of household[10]	Capo famiglia

---

[1] Age at last birthday.

[2] The classification by level of education is by the highest level of educational certification attained. There are five levels as noted.

[3] In the table titles, "industrial sector" (<u>settore di attività economica</u>) designates the following abbreviated industrial classification: (1) agriculture, hunting, fishing; (2) industries (extractive, manufacturing, construction, power); (3) other activities.

[4] In the table titles, "occupation" designates throughout a full tabulation of all three levels of occupational groups.

[5] In the table titles, "occupational/non-occupational status" groups the population according to whether it does or does not have an occupation. Persons "having an occupation" were 10 years of age or older on the census day and were employed, temporarily unemployed but seeking work, or temporarily not engaged in their usual occupation because of military service, imprisonment for fewer than five years, etc. Persons "not having an occupation" were at least 14 years of age and were seeking their first employment or were economically inactive persons, such as students, women doing housework, rentiers, the permanently disabled, and prisoners sentenced for more than five years.

[6] The <u>de facto</u> population: all persons present in a commune on the census day habitually residing there, as well as all other

persons present in the commune on the census day habitually residing in another commune or abroad.

[7] The de jure population: all persons present in a commune on the census day habitually residing there, as well as all those habitually residing in the commune who were absent on the census day because of military draft, temporary placement in a nursing home, or the like.

[8] In the table titles, "private household" designates a group of persons bound by ties of consanguinity, affinity, adoption, affection, etc., who habitually reside in the same dwelling and who fulfill their common needs through joint disposition of their income. Households are classified into four types:

- A. The head of the household with or without unrelated members.
- B. The head of the household and his spouse with or without unrelated members.
- C. The head of the household with or without spouse, with children, and with or without unrelated members.
- D. The head of the household with or without his spouse, with or without children, with other relatives, and with or without related members.

[9] In the table titles, "institutional household" designates a group of persons who are not bound by family ties but who live together for religious, health, military, penal, or similar reasons. The category also includes persons living together as a result of employment or work at an institution, provided they do not constitute a family as such.

[10] In the table titles, "head of household" designates the person who exercises "paternal authority" by managing family affairs and taking care of family interests.

TABLE TITLES BY VOLUME

VOL. I. SUMMARY DATA BY COMMUNE AND PROVINCE ON POPULATION AND HOUSING.

TABLES

Number	Title	Page
1.	Resident and enumerated population by province.	9
	A. Total province.	9
	B. Chief town of each commune.	11
	C. Remainder of each commune.	13

TABLE TITLES BY VOLUME

2. Resident population by province and statistical section.[1] — 15
3. Resident population by province, altimetric zone, and rural and urban communes. — 22
4. Communes by province and commune size-class. — 26
5. Members of private and institutional households by province. — 30
6. Housing by province. — 32
   A. Total province. — 32
   B. Chief town of each commune. — 34
   C. Remainder of each commune. — 36
7. Resident population, enumerated population, and housing by commune. — 38

VOL. II. SUMMARY DATA BY COMMUNE AND PROVINCE ON SOME PRINCIPAL STRUCTURAL CHARACTERISTICS OF THE POPULATION: SEX, AGE, LEVEL OF EDUCATION, AND ECONOMIC ACTIVITY.

TABLES

Number	Title	Page
	National Data	
1.	Resident population by sex and age group.	9
2.	Resident population 6 years of age or more by sex and level of education.	9
3.	Economically active and inactive resident population by sex; economically active population by industrial sector.	9
	Provincial Data	
4.	Resident population by sex and age group.	10
5.	Resident population 6 years of age or more by sex, age group, and level of education.	16
6.	Resident population 10 years of age or more, economically active and inactive, by sex and age group; economically active population by industrial sector.	44
	Communal Data	
7.	Resident population by sex and age group.	72

---

[1] A statistical section is an areal unit intermediate in size between the province and the commune.

ITALY

8. Resident population 6 years of age or more by sex and level of education.   300

9. Resident population 10 years of age or more, economically active and inactive, by sex; economically active population by industrial sector.   416

VOL. III. SUMMARY DATA BY COMMUNE.
[92 fascicles; see listing on pp. 218-220 above.]

Structure of the Volume

The table numbers and table titles for all fascicles of Volume 3 are identical, except for Fascicles 21 and 32,[1] each of which contains eight additional table titles for data arranged by language spoken. Page locations vary from fascicle to fascicle. Hence, it is necessary to consult the Tables list as well as the Page Location Matrix which follow in order to locate desired data.

The procedure for obtaining page locations of specific data involves three steps:

1. Consult Tables list for desired data and table number.
2. Consult the Fascicle notes (pp. 218-220 above) for desired commune and number of pages in the fascicle. The number of pages represents the Index number to be used in the Page Location Matrix. In Fascicles 9, 21, 32, and 88, the Index number does not correctly indicate the page locations. Hence, the page locations in these fascicles have been presented separately.
3. Using the Index number and table number, consult the Page Location Matrix for page indicated for data desired.

TABLES

Number	Title
1.	Geographical area and density of the commune. Resident and enumerated population by sex; resident population by kind of populated areas; resident population temporarily absent by current visiting place.
2.	Resident population in communes, geographical sections, and inhabited areas by altitude.
3.	Resident population by sex and marital status.

---

[1] These fascicles appear at the end of the Page Location Matrix.

TABLE TITLES BY VOLUME

4. Resident population by sex and age group.

5. Resident population 6 years of age or more by sex and level of education.

6. Resident population 10 years of age or more economically active and inactive, by sex; economically active persons by industrial division; economically inactive persons by classification of persons not having an occupation.

7. Resident population having an occupation by sex, industrial sector, and employment status.

8. Private households (resident population) by industrial division or non-occupational status of the head of the household.

9. Private households (resident population) with the head of the household having an occupation by industrial sector and employment status of the head.

10. Occupied and unoccupied dwellings by kind of tenure.

11. Occupied and unoccupied dwellings by facilities installed.

12. Resident population by sex and language spoken; resident population temporarily absent by current visiting place.

13. Resident population by sex, marital status, and language spoken.

14. Resident population by sex, age group, and language spoken.

15. Resident population 6 years of age and older by sex, level of education, and language spoken.

16. Resident population 10 years of age and older, economically active and inactive, by sex and language spoken; economically active persons by industrial division; economically inactive persons by classification of persons not having an occupation.

17. Resident population having an occupation by sex, language spoken, industrial sector, and employment status.

18. Occupied dwellings by language spoken by the head of the household and by kind of tenure.

19. Occupied dwellings by language spoken by the head of the household and by facilities installed.

ITALY

PAGE LOCATION MATRIX

Fascicle Numbers	Index	Table Numbers										
		1	2	3	4	5	6	7	8	9	10	11
		Pages										
74, 86	25	12	12	14	15	16	16	18	20	20	22	24
88	25	12	12	14	14	16	16	18	20	20	22	24
31, 49, 73, 85, 89	27	12	12	16	16	18	18	20	22	22	24	26
39	29	12	12	18	18	20	20	22	24	24	26	28
81	31	12	12	20	20	22	22	24	26	26	28	30
45, 47, 53	33	12	12	22	22	24	24	26	28	28	30	32
38	35	12	12	24	24	26	26	28	30	30	32	34
77	37	12	14	16	17	20	22	24	28	30	34	36
59, 72, 84	39	12	14	18	19	22	24	26	30	32	36	38
56, 68, 71, 87	41	12	14	20	21	24	26	28	32	34	38	40
8, 11, 55	43	12	14	22	23	26	28	30	34	36	40	42
7, 27, 29, 40, 44, 50, 52, 67	45	12	14	24	25	28	30	32	36	38	42	44
9	45	12	14	22	23	26	28	30	34	36	40	42

TABLE TITLES BY VOLUME

Fascicle Numbers	Index	\multicolumn{11}{c}{Table Numbers}
		1	2	3	4	5	6	7	8	9	10	11
		\multicolumn{11}{c}{Pages}
42, 43, 57	47	12	14	26	27	30	32	34	38	40	44	46
10, 35, 41	49	12	14	28	29	32	34	36	40	42	46	48
20, 25, 51	51	12	14	30	31	34	36	38	42	44	48	50
33, 36, 37, 46, 48	53	12	14	32	33	36	38	40	44	46	50	52
34	57	12	14	36	37	40	42	44	48	50	54	56
54	59	12	14	38	39	42	44	46	50	52	56	58
75, 90, 91	65	12	16	22	24	30	34	38	46	50	58	62
62, 76, 82	67	12	16	24	26	32	36	40	48	52	60	64
14, 61, 63, 64, 69, 70	69	12	16	26	28	34	38	42	50	54	62	66
12, 58, 60, 80	71	12	16	28	30	36	40	44	52	56	64	68
5, 66, 83	73	12	16	30	32	38	42	46	54	58	66	70
19, 26, 28	75	12	16	32	34	40	44	48	56	60	68	72
23	81	12	16	38	40	46	50	54	62	66	74	78
24	83	12	16	40	42	48	52	56	64	68	76	80
92	95	12	18	30	33	42	48	54	66	72	84	90
79	97	12	18	32	35	44	50	56	68	74	86	92
78	99	12	18	34	37	46	52	58	70	76	88	94
65	101	12	18	36	39	48	54	60	72	78	90	96

Fascicle Numbers	Index	Table Numbers										
		1	2	3	4	5	6	7	8	9	10	11
						Pages						
2, 3	103	12	18	38	41	50	56	62	74	80	92	98
6	105	12	18	40	43	52	58	64	76	82	94	100
18	109	12	18	44	47	56	62	68	80	86	98	104
30	119	12	18	54	57	66	72	78	90	96	108	114
13	131	12	20	44	48	60	68	76	92	100	116	124
16, 17, 22	133	12	20	46	50	62	70	78	94	102	118	126
15	137	12	20	50	54	66	74	82	98	106	122	130
4	143	12	20	56	60	72	80	88	104	112	128	136
1	171	12	22	62	67	82	92	102	122	132	152	162

Fascicle Numbers	Index	Table Numbers: Fascicles 21 and 32																	
		1	2	3	4	5	6	7	8	9	10	11	12	13	14	15	16	17	18	19
												Pages								
32	31	12	12	15	15	16	16	16	18	18	20	20	24	24	25	26	28	28	31	31
21	229	12	16	34	36	42	46	50	58	62	70	74	80	90	100	130	150	170	210	220

TABLE TITLES BY VOLUME

VOL. III.   SUMMARY DATA BY COMMUNE.   APPENDIX: SUMMARIZED NATIONAL DATA.

TABLES

Number	Title	Page
1.	Geographical areas and density of provinces and regions. Resident and enumerated population by sex; resident population by kind of populated areas; resident population temporarily absent by current visiting place.	12
2.	Resident population by sex and marital status.	16
3.	Resident population by sex and age group.	18
4.	Resident population 6 years of age or more by sex and level of education.	24
5.	Economically active and inactive population 10 years of age or more by sex; economically active by industrial division; economically inactive by classification of persons not having an occupation.	28
6.	Economically active resident population having an occupation by sex, industrial sector, and employment status.	32
7.	Private households (resident population) by the head of the household's industrial division, or by his position in the classification of persons not having an occupation.	40
8.	Private households (resident population), the heads of which have an occupation, by industrial sector and employment status of the head of the household.	44
9.	Occupied and unoccupied housing by kind of tenure.	52
10.	Occupied and unoccupied housing by facilities installed.	56

VOL. IV.   HOUSEHOLDS AND INSTITUTIONS.

TABLES

Number	Title	Page
	Private Households: National and Regional Data	
1.	Private households (resident population) by type and size of household.	11
2.	Private households (resident population) and members by type and size of household.	11
3.	Private households (resident population) and members	

ITALY

	by type of household and occupational/non-occupational status of the head of the household.	14
4.	Private households (resident population) by size of household and sex, marital status, and age group of the head of the household.	24
5.	Private households (resident population) by size of household and occupational/non-occupational status of the head of the household, by region.	26
6.	Private households (resident population) whose heads have an occupation by size of household and age group, industrial sector, and employment status of the head of the household.	66
7.	Private households (resident population) by size of household and occupational/non-occupational status and educational level of the head of the household, by region.	76
8.	Private households (resident population) by size of household and number of household members having an occupation, by region.	96
9.	Private households (resident population) whose heads have an occupation by size of household, number of household members having an occupation, and industrial sector and employment status of the head of the household.	116
10.	Private households (resident population) by size of household and occupational/non-occupational status of household members, by region.	120
11.	Private households (resident population) and members by occupational/non-occupational status of the head of the household, and of relatives (e.g., grandparents), and of in-laws, by region.	160
12.	Private households (resident population) by number of children living in the household and sex, marital status, and age group of the head of the household.	200
13.	Children living in the household by sex and age group, and by occupational/non-occupational status of the head of the household, by region.	202
14.	Children living in the household by sex, age group, and occupational/non-occupational status, and by occupational/non-occupational status of the head of the household.	212
15.	Private households (resident population) by number of children living in the household and number of children having an occupation and by occupational/non-occupational status of the head of the household, by region.	218

TABLE TITLES BY VOLUME

16. Private households (resident population) by type of household, by province. — 260
17. Private households (resident population) by size and type of household, by province. — 266
18. Private households (resident population) and members by type of household, by province. — 276
19. Private households (resident population) by size of household and sex and marital status of the head of the household, by province. — 286
20. Private households (resident population) by size of household and occupational/non-occupational status of the head of the household, by province. — 301
21. Private households (resident population) by occupational/non-occupational status and level of education of the head of the household, by province. — 412
22. Private households (resident population) by occupational/non-occupational status of the head of the household, total number of household members, and number of household members having an occupation, by province. — 430
23. Private households (resident population) whose heads have an occupation by industrial sector, total number of household members, and number of household members having an occupation, by province. — 486
24. Private households (resident population) by occupational/non-occupational status of household members and of the head of the household, by province. — 542
25. Private households (resident population) by number of children living in the household and sex and marital status of the head of the household, by province. — 570
26. Private households (resident population) by number of children living in the household and occupational/non-occupational status of the head of the household, by province. — 584
27. Children living in the household by sex and age group and by occupational/non-occupational status of the head of the household, by province. — 598
28. Children living in the household by age group and occupational/non-occupational status of the children and of the head of the household, by province. — 612

Institutions

29. Institutions and inmates, by type of institution and province. — 642

ITALY

Number	Title	Page
30.	Inmates in institutions by sex, type of institution, and province.	646
31.	Institutions by type, number of inmates by sex, and province.	654

VOL. V. SEX, AGE, MARITAL STATUS, PLACE OF BIRTH.

TABLES

Number	Title	Page
1.	Resident and enumerated population by age and marital status.	8
	A. Nation.	8
	B. Regions.	9
	C. Provinces.	28
2.	Resident population by sex, age, and marital status.	120
	A. Nation.	120
	B. Regions.	121
	C. Provinces.	140
3.	Enumerated population by sex, age, and marital status.	232
	A. Nation.	232
	B. Regions.	233
	C. Provinces.	252
4.	Resident population by sex, year of birth, and marital status.	344
	A. Nation.	344
	B. Regions.	348
5.	Enumerated population by sex, year of birth, and marital status.	424
	A. Nation.	424
	B. Regions.	428
6.	Resident population by resident province and place of birth.	504
7.	Resident population by sex, age group, and place of birth.	508
	A. Nation.	508
	B. Regions.	512
	C. Provinces.	550

TABLE TITLES BY VOLUME

VOL. VI. OCCUPATIONS.

TABLES

Number	Title	Page
	National Data	
1.	Economically active and inactive resident population by sex.	10
2.	Economically active and inactive resident population by sex, age, and marital status.	11
3.	Economically active resident population by sex, age group, marital status, and industrial division.	14
4.	Economically active resident population having an occupation by sex, age group, marital status, and employment status.	16
5.	Economically active resident population having an occupation by sex and occupation.	18
6.	Economically active resident population having an occupation by sex, age group, marital status, and occupation.	24
7.	Economically active resident population having an occupation by sex, occupation, employment status, and industrial division.	48
8.	Economically active resident population having an occupation by sex, employment status, and industrial division (major group, group, and category).	72
9.	Economically inactive resident population by sex, age group, marital status, and classification of persons not having an occupation.	88
	Provincial and Regional Data	
10.	Economically active and inactive resident population by sex, by province.	92
11.	Economically active resident population by sex, marital status, and industrial division, by province.	94
12.	Economically active resident population by sex, age group, and industrial division, by province.	110
13.	Economically active resident population having an occupation by sex and employment status, by province.	166
14.	Economically active resident population having an occupation by sex, age group, and employment status, by province.	170
15.	Economically active resident population having an occupation by sex and occupation, by province.	194

ITALY

16.	Economically active resident population having an occupation by sex and industrial division (major group, group, and subgroup), by region.	356
17.	Economically active resident population having an occupation by sex, occupational status, industrial division, and major group.	420
	A. Regions	420
	B. Provinces	458
18.	Economically active resident population having an occupation by sex, occupation, employment status, and industrial division, by region.	640
19.	Economically inactive resident population by sex and classification of persons not having an occupation, by province.	1096
20.	Economically inactive resident population by sex, age group, and classification of persons not having an occupation, by province.	1100
21.	Economically inactive resident population by sex, marital status, and classification of persons not having an occupation, by province.	1128
22.	Economically active resident population of institutions by industrial status of the institution[1] and sex, by province.	1136
23.	Economically inactive resident population of institutions by classification of persons not having an occupation, by province.	1144

VOL. VII. EDUCATION.

TABLES

Number	Title	Page
	National Data	
1.	Resident population 6 years of age or more by sex, age group, and level of education.	11
2.	Economically active and inactive resident population 6 years of age or more by sex and level of education.	14
3.	Economically active and inactive resident population 6 years of age or more by sex, age group, and level of education.	20

---

[1] The industrial status of the institution refers to the type of service performed (e.g., public administration).

4.	Economically active resident population having an occupation by sex, industrial division, employment status, and level of education.	26
5.	Economically active resident population having an occupation by sex, occupation, employment status, and level of education.	36
6.	University graduates having an occupation by sex, kind of degree, and occupation.	50
7.	Diplomates having an occupation by sex, kind of diploma, and occupation.	58
8.	Resident population 14 years of age or more who are in search of their first job by sex, age group, and level of education.	66

Provincial Data

9.	Resident population 6 years of age or more by level of education and province.	68
10.	Resident population 6 years of age or more by sex, age group, level of education, and province.	72
11.	Economically active and inactive resident population 6 years of age or more by sex, level of education, and province.	402
12.	Economically active resident population having an occupation by sex, age group, industrial sector, level of education, and province.	416
13.	Economically active resident population having an occupation by sex, industrial division, employment status, level of education, and province.	472
14.	Resident population 14 years of age or more who are in search of their first job by sex, age group, level of education, and province.	583
15.	Students and school children 6 years of age or more by sex, age group, level of education, and province.	596
16.	Economically inactive resident population 6 years of age or more by sex, age group, level of education, and province.	608

VOL. IX. GENERAL SUMMARY DATA.

TABLES

Number	Title	Page
	Population and Areal Units	
1.	Geographical areas and density of provinces and	

ITALY

	regions. Resident and enumerated population by sex; resident population by kind of populated areas; resident population temporarily absent by current visiting place.	16
2.	Resident population by altimetric zone, degree of urbanization or rurality of the communes, and provinces.	20
3.	Communes by geographical size and province.	24
4.	Communes by altitude and province.	28
5.	Communes by commune size-class, altimetric zone, and province.	32
	A. Total.	32
	B. Mountains.	36
	C. Hills.	40
	D. Plains.	44
6.	Populated and unpopulated geographical sections by province; populated sections by size-class, and unpopulated sections by type.[1]	48
7.	Population centers[2] and unpopulated areas by province; population centers by size-class.	52
8.	Population nuclei[3] and unpopulated areas by province; population nuclei by size-class.	56

Private and Institutional Households

9.	Members of private and institutional households by province.	63
10.	Private households (resident population) by type and size of household.	65
11.	Private households (resident population) and members by size of household.	65

---

[1] Unpopulated sections are tabulated by (1) high mountain areas and (2) lakes, swamps.

[2] A population center is an agglomeration of contiguous or neighboring houses with roads, squares, and public facilities and services where meetings for business, education, religion, etc. can be held.

[3] A population nucleus is an agglomeration of houses with at least 5 households but lacking the public facilities and services which distinguish a population center as a meeting place for business, etc. Nuclei may exist in remote areas with difficult access by road; they may consist of ruined, no longer inhabited houses in high mountain areas; and they include convents, sanatoriums, orphanages, etc. located in open expanses of land.

12.	Private households (resident population) by size of household and by sex, marital status, and age group of the head of the household.	66
13.	Private households (resident population) and members by occupational/non-occupational status of the head of the household.	68
14.	Private households (resident population) by size of household and the occupational/non-occupational status of the head of the household. A. Nation. B. Regions.	70 70 71
15.	Private households (resident population) by size of household and number of members having an occupation.	90
16.	Private households (resident population) whose head has an occupation by size of household, number of members having an occupation, and employment status of the head of the household.	91
17.	Private households (resident population) by size of household and occupational/non-occupational status of the members of the household.	92
18.	Private households (resident population) by number of children in the household, occupational/non-occupational status of the head of the household, and province.	94
19.	Institutional households and their permanent and temporary members by kind of institution[4] and province.	108

Sex, Age, Marital Status, and Place of Birth

20.	Resident and enumerated population by age and marital status. A. Nation. B. Regions.	114 114 115
21.	Resident population by sex, age, and marital status. A. Nation. B. Regions.	134 134 135
22.	Enumerated population by sex, age, and marital status. A. Nation. B. Regions.	154 154 155
23.	Resident population by sex, age, and province.	174

---

[4] The following kinds of institutions are tabulated: educational, religious, charitable, treatment, penal, lodging, ships, and other institutions.

ITALY

24.	Enumerated population by sex, age, and province.	187
25.	Resident population by sex, year of birth, and marital status.	200
26.	Enumerated population by sex, year of birth, and marital status.	204
27.	Resident population by province of residence and place of birth.	208

Occupations

28.	Economically active and inactive resident population by sex, age, and marital status.	215
29.	Economically active resident population by sex, age group, marital status, and industrial division.	218
30.	Economically active resident population having an occupation by sex, age group, marital status, and employment status.	220
31.	Economically active resident population having an occupation by sex, age group, and occupation.	222
32.	Economically active resident population having an occupation by sex, occupation, employment status, and industrial division.	232
33.	Economically inactive resident population by sex, age group, marital status, and classification of persons not having an occupation.	256
34.	Economically active and inactive resident population by sex and province.	258
35.	Economically active resident population having an occupation by sex, occupation, and region.	260
36.	Economically active resident population having an occupation by sex, employment status, industrial division and major group, and region.	296
37.	Economically active resident population by sex, industrial division, and province.	334
38.	Economically active resident population having an occupation by sex, employment status, and province.	338
39.	Economically inactive resident population by sex, classification of persons not having an occupation, and province.	342
40.	Resident population in institutions: economically active by occupation;[5] economically inactive by classification of persons not having an occupation.	346

---

[5] The classification by occupation is limited to the following

TABLE TITLES BY VOLUME

Education

41. Resident population 6 years of age or more by sex, age group, and level of education. 353
42. Economically active and inactive resident population 6 years of age or more by sex and level of education. 356
43. Economically active resident population having an occupation by sex, employment status, and level of education. 362
44. Economically active resident population having an occupation by sex, industrial division, employment status, and level of education. 363
45. Resident population 6 years of age or more by sex, level of education, and province. 364
46. Economically active resident population having an occupation by sex, industrial division, level of education, and province. 376
47. Resident population 14 years of age or more who are in search of their first job by sex, level of education, and province. 398
48. Economically inactive resident population 6 years of age or more by sex, level of education, and province. 400

Housing

49. Occupied and unoccupied housing by kind of tenure and number of rooms. 415
50. Occupied and unoccupied housing by facilities installed. 415
51. Occupied and unoccupied housing and other occupied lodgings by province. 416
52. Occupied and unoccupied housing by kind of tenure and province. 420
53. Occupied and unoccupied housing by number of rooms and province. 424
54. Occupied and unoccupied housing by facilities installed and province. 426
55. Occupied housing by kind of tenure and occupational/non-occupational status of the head of the household;

---

categories: (1) various social services and activities, (2) public administration, (3) other occupations, (4) those in search of their first job, and (5) total occupations.

ITALY

	other occupied lodgings by occupational/non-occupational status of the head of the household.	428
56.	Occupied housing by number of rooms and occupational/non-occupational status of the head of the household.	430
57.	Occupied housing by facilities installed and occupational/non-occupational status of the head of the household.	431
58.	Occupied housing by level of occupancy[6] and occupational/non-occupational status of the head of the household.	432
59.	Occupied housing by level of occupancy[6] and province.	434
60.	Occupied housing by number of rooms, number of occupants, and province.	438

Persons Temporarily Absent

61.	Persons temporarily absent from their commune of residence by sex, reason for absence,[7] current visiting place, and province of residence.	458
62.	Persons temporarily absent from their commune of residence but present in another Italian commune by sex, reason for absence,[7] region where present, and province of residence.	470
63.	Persons temporarily abroad by sex, reason for absence,[7] foreign country visiting,[8] and province of residence.	482

Foreigners

64.	Foreigners resident in Italy by sex, age group, and citizenship.	496
	A. Total.	496
	B. Males.	498
	C. Females.	500

---

[6] Three levels of occupancy are tabulated: (1) not crowded: no more than one person per room; (2) crowded: between one and two persons per room; and (3) overcrowded: more than two persons per room.

[7] The reason for absence is tabulated according to work-related or other reasons.

[8] The following countries are tabulated: France, the Federal Republic of Germany, Belgium, the Netherlands, Luxemburg, the United Kingdom, Switzerland, and other European countries. Tabulations are also given for Africa, the Americas, Asia, Oceania, and persons at sea.

65. Foreigners resident in Italy by sex, citizenship, and Italian province of residence. 502

66. Economically active and inactive foreigners resident in Italy by sex and citizenship. 520

67. Foreigners resident in Italy having an occupation by sex, occupation, and employment status. 522

TENTH GENERAL CENSUS OF THE POPULATION, 15 OCTOBER 1961. RESIDENT POPULATION BY COMMUNES.

Structure of the Volume

This volume consists of tabulations of the resident (legal) population of communes grouped within their respective provinces, which have been listed alphabetically.

Province	Page	Province	Page
Agrigento	6	Forlì	80
Alessandria	7	Frosinone	81
Ancona	11	Genova	83
Aosta (Valle di)	12	Gorizia	85
Arezzo	14	Grosseto	85
Ascoli Piceno	15	Imperia	86
Asti	17	L'Aquila	88
Avellino	19	La Spezia	90
Bari	22	Latina	91
Belluno	23	Lecce	92
Benevento	25	Livorno	94
Bergamo	27	Lucca	95
Bologna	32	Macerata	95
Bolzano--Bozen	34	Mantova	97
Brescia	36	Massa-Carrara	98
Brindisi	41	Matera	99
Cagliari	41	Messina	100
Caltanissetta	45	Milano	102
Campobasso	46	Modena	107
Caserta	49	Napoli	109
Catania	51	Novara	111
Catanzaro	53	Nuoro	114
Chieti	56	Padova	117
Como	59	Palermo	119
Cosenza	64	Parma	121
Cremona	68	Pavia	122
Cuneo	70	Perugia	126
Enna	76	Pesaro e Urbino	128
Ferrara	76	Pescara	129
Firenze	77	Piacenza	130
Foggia	78	Pisa	131

ITALY

Province	Page	Province	Page
Pistoia	132	Taranto	155
Potenza	133	Teramo	156
Ragusa	135	Terni	157
Ravenna	136	Torino	157
Reggio di Calabria	136	Trapani	164
Reggio nell'Emilia	138	Trento	165
Rieti	139	Treviso	170
Roma	141	Trieste	172
Rovigo	144	Udine	172
Salerno	145	Varese	177
Sassari	148	Venezia	180
Savona	150	Vercelli	181
Siena	152	Verona	185
Siracusa	153	Vicenza	187
Sondrio	153	Viterbo	190

L I E C H T E N S T E I N

MAIN ENTRY:

Liechtenstein. Amt für Statistik.
    Liechtensteinische Volkszählung, 1. Dezember 1960. [Vaduz, 1963]
    1 v. (unpaged)   30cm.   (65-72350)

ENGLISH TRANSLATION OF CENSUS TITLE:

Liechtenstein population census, December 1, 1960.

ENUMERATION DATE: December 1, 1960.

ENUMERATION BASIS: De facto; de jure.

GLOSSARY

ADMINISTRATIVE, GEOGRAPHICAL, AND POLITICAL UNITS

Level	English Translation	Vernacular Term
National	Principality of Liechtenstein	Fürstentum Liechtenstein
Intermediate	...	...
Local	Commune	Gemeinde
Other concept:		
	Commune size-group	Gemeindegruppe

    The __Principality of Liechtenstein__ is composed of 11 __communes__. Communes are grouped into __commune size-groups__ ranging in population size from 0-999 to 2000-4999.

TABLE OF DEMOGRAPHIC CONCEPTS

    The demographic concepts which appear in the following table correspond to the recommendations of the European Conference of Statisticians, unless otherwise noted. They are defined and discussed in the appendixes and in footnotes. Terms in parentheses are alternative forms which appear in the table titles. Terms in brackets are not used in the table titles, but they are included here in order to give logical completeness to the various classifications in the table.

Appendix	English Table Term	Vernacular Table Term
	ECONOMIC CHARACTERISTICS	
A	Economic activity	Erwerbsleben
	Economically active	Aktive Bevölkerung (Berufstätige)
	Economically inactive	Nicht aktive Bevölkerung
B	Employment status	Beruflicher Stellung
C	Industrial classification[1]	Erwerb
	[Division][2]	...
	Major group(s)[3]	Erwerbsgruppe
	Group(s)[4]	Erwerbsart
D	Occupational classification	Beruf
	[Major group][5]	...
	[Minor group][6]	...
	Unit group(s)[7]	Personlicher Beruf
	HOUSEHOLD AND FAMILY CHARACTERISTICS	
F	Household	Haushaltung
	Private	Privathaushalt
	Institutional	Anstalt

[1] In the table titles, "industrial classes" (Erwerbsklassen) designates the following abbreviated industrial classification: (1) agriculture, forestry; (2) industry, manufacturing, construction; (3) trade, services, tourism.

[2] In the table titles, the term "industrial division" does not appear. However, tabulations of this level are always provided along with those of major groups and groups.

[3] In the table titles, "industrial major group" designates tabulations of industrial divisions and major groups.

[4] In the table titles, "industrial group" designates tabulations of industrial divisions, major groups, and groups.

[5] Tabulations of occupational major groups are not given.

[6] In the table titles, the term "industrial minor group" does not appear. However, tabulations of this level are always provided along with those of occupational unit groups.

[7] In the table titles, "occupational unit group" designates tabulations of occupational minor and unit groups.

TABLE TITLES

LIECHTENSTEIN POPULATION CENSUS, DECEMBER 1, 1960.
TABLES[1]

Number	Title
1.	All residents, Liechtensteiners, and foreigners by marital status, birthplace, place of origin,[2] religious affiliation, mother tongue, age group, economic dependence,[3] and size-group of commune.
2.	Households by commune, residents by commune, sex, marital status, birthplace, place of origin, religious affiliation and mother tongue.
3.	Residents by commune, age group, and sex; economically active population by commune, industrial class, and sex; total, Liechtensteiners, foreigners.
4.	Inhabited buildings and residences by commune.
5.	Foreigners living in the Principality of Liechtenstein by nationality.
6.	Inhabitants of the Principality of Liechtenstein born abroad by birthplace.
7.	Residents by year of birth, sex, place of origin, and marital status.
8.	Residents by five-year age group, sex, place of origin, and marital status.
9.	Economically active persons and their economically inactive dependents by industrial group.
9a.	Staff of institutions, their economically inactive dependents, and inmates of institutions by kind of institution.
10.	Economically active population by industrial major group and employment status.
11.	Economically inactive dependents of economically active persons by industrial major group and employment status of supporter.

---

[1] No page numbers provided in original text.

[2] The place of origin (birth) of persons born in Liechtenstein is tabulated according to a dichotomous classification--present commune of residence/all other communes. For those born outside Liechtenstein, the place of origin (birth) is the birthplace according to present national boundaries.

[3] Economically active and inactive, occupation, industry, employment status, etc.

LIECHTENSTEIN

12. Economically active persons and their economically inactive dependents in private and public sectors of the economy by industrial major group.

13. Economically active persons and their economically inactive dependents by occupational unit group, employment status, sex, place of origin, and marital status; total, Liechtensteiners, foreigners.

14. Economically active population by occupational unit group, employment status, sex, and age.

15. Members of families working together by industrial major group and employment status.

16. Wives by year of marriage, number of live-born children, and age of marriage.

L U X E M B U R G

MAIN ENTRY AND VOLUME NOTES:[1]

Luxemburg. Service Central de la Statistique et des Études Économiques.
  Recensement de la population du 31 décembre 1960. Luxembourg, [1962?]-1968.
    6 v. 30cm.

- I. Population, maisons et ménages au 31.12.1962 par communes, sections, et localités; et tableaux rétrospectifs. [1962?] 85 p.
- II. Caractéristiques personnelles (sexe, âge, état matrimonial, nationalité, lieu de naissance, culte). 1966. 47 p.
- III. Caractéristiques économiques de la population (statut professionnel, activité, profession principale, profession accessoire, catégorie socio-économique). 1967. 124 p.
- IV. Ménages et familles. 1967. 24 p.
- V. Maisons et logements.[2] [Excluded.]
- VI. Méthodologie et législation.[2] [Excluded; no tables.]

ENGLISH TRANSLATION OF CENSUS AND VOLUME TITLES:

Census of the population, December 31, 1960.

- I. Population, housing, and households on 31 December 1962 by commune, electoral division, and locality; and retrospective tables.
- II. Personal characteristics of the population (sex, age, marital status, nationality, birthplace, religion).
- III. Economic characteristics of the population (employment status, economic activity, principal occupation, secondary occupation, socio-economic status).
- IV. Households and families.
- V. Dwellings and houses.

---

[1] The spelling of the word Luxembourg/Luxemburg follows the customary usage of the National Union Catalog of the Library of Congress, where "burg" is used for filing.

[2] Volumes V and VI are published as one, continuously paged: Volume V, pp. 1-33; Volume VI, pp. [35]-103.

LUXEMBURG

VI. Methodology and legal bases.

ENUMERATION DATE: December 31, 1960.

ENUMERATION BASIS: De facto; de jure.

GLOSSARY

ADMINISTRATIVE, GEOGRAPHICAL, AND POLITICAL UNITS

Level	English Translation	Vernacular Term
National	Grand Duchy of Luxembourg	Grand-Duché de Luxembourg
Intermediate	District	District
Local	Canton	Canton
	Commune	Commune
Other concepts:	Electoral constituency	Circonscription électorale
	Electoral division	Section électorale
	Locality	Localité

Districts are the largest administrative subdivisions in Luxemburg. There are three districts, each headed by a centrally appointed commissioner who serves as an intermediate authority between the national government and the local government.

Cantons are subdivisions of the districts. There are a total of 12.

Communes are subdivisions of cantons. The 126 communes are the smallest administrative units in the country.

Electoral constituencies, of which there are four, are geographical divisions of the country: South, East, Central, and North.

Electoral divisions are subdivisions of the communes, and are in turn divided into localities, the smallest areal units in the nation (cities, towns, etc.).

TABLE OF DEMOGRAPHIC CONCEPTS

The demographic concepts which appear in the following table correspond to the recommendations of the European Conference of Statisticians, unless otherwise noted. They are defined and discussed in the appendixes and in footnotes. Terms in parentheses are alternative forms which appear in the table titles.

GLOSSARY

Terms in brackets are not used in the table titles, but they are included here in order to give logical completeness to the various classifications in the table.

Appendix	English Table Term	Vernacular Table Term
	ECONOMIC CHARACTERISTICS	
A	Economic activity	Activité économique
	Economically active[1]	Population active
	Economically inactive	Population non active
B	Employment status	Statut professionnel
C	[Industrial classification][2]	...
	Division(s)	Branche d'activité
	Major group(s)	Classe d'activité
D	[Occupational classification]	...
	Major group(s)	Grands groupes
	Minor group(s)	Sous-groupes
	Unit group(s)	Groupes de base
E	Socio-economic category	Catégorie socio-économique
	HOUSEHOLD AND FAMILY CHARACTERISTICS	
F	[Household]	...
	Private household	Ménage privé
	Institutional household	Ménage collectif
	Head of the household	Chef de ménage
	Family nucleus	Noyau familial
	Marital status[3]	État matrimonial

---

[1] Differs from the definition of the European Conference of Statisticians in that unemployed persons are included with the inactive population as a separate group.

[2] In the table titles, "industrial sector" (secteur économique) designates an abbreviated industrial classification of three categories as follows: agriculture, industry, and services.

[3] Marital status comprises the following categories: single or never married; married (including stable de facto unions); widowed, not remarried; divorced, not remarried; status not given.

LUXEMBURG

TABLE TITLES BY VOLUME

I. POPULATION, HOUSING AND HOUSEHOLDS ON 31 DECEMBER 1962 BY COMMUNE, ELECTORAL DIVISION, AND LOCALITY; AND RETROSPECTIVE TABLES.

TABLES

Number	Title	Page
	(A) Results of the 1960 Census	
I.	Results by district, canton, and electoral constituency.	9
II.	Results by canton and commune.	10
III.	Cantons and communes by size of population.	13
IV.	Cantons and communes by density[1] of population.	15
V.	Results by commune, electoral division, and locality.	16
VI.	Alphabetical list of communes.	47
VII.	Alphabetical list of localities.	50
	(B) Retrospective Tables	
I.	The population of the Grand Duchy, 1821 to 1960.	71
II.	The population, by commune and canton, 1821 to 1960.	74
III.	The number of houses and the number of households, 1864 to 1960.	79
IV.	The number of inhabited dwellings by commune and canton, 1871 to 1960.	80
V.	The number of households by commune and canton, 1871 to 1960.	83

II. PERSONAL CHARACTERISTICS OF THE POPULATION (SEX, AGE, MARITAL STATUS, NATIONALITY, BIRTHPLACE, RELIGION).

TABLES

Number	Title	Page
	(A) Results of the Population Census of 31-12-60 for the Entire Country	
201.	Total population by commune size-class and sex.	6
202.	Total population, Luxembourg citizens, and foreigners by age group, sex, and urban or rural residence.	7
203.	Luxembourg citizens and foreigners by age group and sex.	8

---
[1] Inhabitants per square kilometer.

204.	Total population, Luxembourg citizens, and foreigners by single year of age and sex.	10
205.	Total population, Luxembourg citizens, and foreigners by age group, sex, and marital status.	12
206.	Urban and rural populations by age group, sex, and marital status.	13
207.	Total population by age group, sex, and birthplace.	14
208.	Persons of foreign birth by age group, country of birth, and sex.	15
209.	Persons of foreign birth by country of birth and sex, with an indication of the number of Luxembourg citizens of foreign birth.	17
210.	Foreign population and foreigners by principal nationalities, and by age group and sex.	18
211.	Foreign population and foreigners by principal nationalities, duration of residence in the Grand Duchy, and sex.	20
212.	Foreign population by nationality and sex, with an indication of the number of foreigners born in the country.	21

(B) Results of the Population Census of 31-12-60 for Territorial Subdivisions

251.	Total population, Luxembourg citizens, and foreigners by district, canton, and commune, by sex.	26
252.	Total population by district, canton, and commune, by place of birth.	30
253.	Foreign population and foreigners by principal nationalities, by district, canton, and commune, by sex.	33

(C) Population Censuses: Comparative Tables

291.	Total population by sex and marital status, 1880 to 1960.	38
292.	Total population, Luxembourg citizens, and foreigners by sex, 1871 to 1960.	39
293.	Total population by sex and age group, 1880 to 1960.	40
294.	Total population by nationality, 1880 to 1960.	43
295.	Total population by country of birth, 1880 to 1960.	44
296.	Total population, Luxembourg citizens, and foreign population by religion, 1871 to 1960.	45
296a.	Total population, Luxembourg citizens, and foreign population by religion, 1947 and 1960.	46

LUXEMBURG

| 297. | Luxembourg citizens by the way in which they became a citizen, 1935, 1947, 1960. | 47 |
| 298. | Comparison between the population of Luxembourg and of neighboring countries. | 47 |

III. ECONOMIC CHARACTERISTICS OF THE POPULATION (EMPLOYMENT STATUS, ECONOMIC ACTIVITY, PRINCIPAL OCCUPATION, SECONDARY OCCUPATION, SOCIO-ECONOMIC STATUS).

TABLES

Number	Title	Page
	(A) General Structure of the Population with Regard to Economic Activity	
301.	Total population, Luxembourg population, and foreign population by type of economic activity and sex.	8
302.	Total population by type of economic activity, sex, and age.	9
303.	Total population, economically active and inactive population, by age group, sex, and marital status.	11
304.	Economically inactive persons, inactive independent persons, and inmates of institutions by age group, sex, and marital status.	12
305.	Economically active population--total, Luxembourg citizens, foreign--by age group and sex.	13
306.	Total population by socio-economic status, age, and sex.	14
	(B) Economic Characteristics of the Population-- National Results	
	(a) Employment Status of Persons Enumerated	
311.	Economically active population by employment status, sex, and age.	18
312.	Economically active population by employment status, sex, and industrial division.	18
313.	Economically active population by employment status, sex, and occupational major group.	19
	(b) Activity of the Firm or Agency Where the Enumerated Person Is Employed	
321.	Economically active population--total, Luxembourg citizens, foreign--by industrial division and sex.	22
322.	Economically active population by industrial division and major group, sex, and age.	23

323.	Economically active population by industrial division and major group, sex, and employment status.	26
324.	Economically active population by industrial group and sex.	28
325.	Economically active and inactive independent population who are dependent on each industrial division by industrial division and employment status, with number of foreigners noted.	33
326.	Economically active and inactive independent population who are dependent on each industrial division and major group by industrial division and major group, and number of heads of households in the economically active population.	34
327.	Economically active population by industrial division, sex, and work place.	36
328.	Economically active population by industrial major group and work place.	37
329.	Economically active foreign population, economically active foreigners from principal nations, by industrial division and major group, sex, and nationality.	39

(c) Principal Occupation

331.	Economically active population--total, Luxembourg citizens, foreign--by occupation (major groups) and sex.	43
332.	Economically active population by occupation (major groups and groups), sex, and age.	44
333.	Economically active population by occupation (major groups and groups), sex, and marital status.	48
334.	Economically active population by occupation (major groups and groups), sex, and employment status.	50
335.	Economically active females by occupation (major groups) and marital status.	53
336.	Economically active population by occupation (unit groups) and sex.	54
337.	Married men whose wives are economically active by socio-economic category and occupation (major groups) of the wife.	62

(d) Secondary Occupation

341.	Persons with a secondary occupation by occupation (major groups and groups), employment status in the principal occupation, and sex.	64
342.	Persons with a secondary occupation by secondary	

	occupation (major groups and groups), employment status in the secondary occupation, and sex.	66
343.	Persons with a secondary occupation by principal occupation (major groups), secondary occupation (major groups), and sex.	68

    (e) Socio-economic Status
[See tables nos. 306 and 337.]

(C) Economic Characteristics of the Population--Results by Territorial Subdivisions

    (a) Activity of the Firm or Agency Where the Enumerated Person Is Employed

361.	Economically active population by district, canton, and commune of residence, industrial division, and sex.	72
362.	Economically active foreign population by district and canton of residence, industrial division, and sex.	82

    (b) Occupation

363.	Economically active population by district, canton and commune of residence, occupation (major groups), and sex.	84

    (c) Work Place and Geographic Mobility of the Labor Force

364.	Economically active population by district, canton, and commune of residence, work place, and sex, with the percentage of outgoing commuters to economically active population in each territorial subdivision.	96
365.	Economically active population by district, canton, and commune of work, residence, and sex, with the percentage of incoming commuters to the labor force working in each territorial subdivision.	101
366.	Outgoing commuters by residence, distance traveled, and means of transportation.	106
367.	Incoming commuters by work place, distance traveled, and means of transportation.	110
368.	Incoming commuters to the capital and to the communes of Esch-sur-Alzette and Schifflange by district, canton, and commune of residence.	115
369.	Incoming commuters to the communes of Differdange, Dudelange, Pétange, and Berg by district and canton of residence.	118

(D) Retrospective Tables

371.	Total population by type of economic activity and sex (1871 to 1960).	120
372.	Total population by socio-economic category and sex (1947 and 1960).	120
373.	Total population and economically active population (agricultural and nonagricultural) by canton of residence (1907 to 1960).	121
374.	Economically active and inactive independent population who are dependent on each industrial sector by sex and industrial sector (1871 to 1960).	122
375.	Economically active population by industrial sector, employment status, and sex (1907 to 1960).	123
376.	Economically active population--Luxembourg citizens and foreign--by industrial sector (1907 to 1960).	124
377.	Economically active population by age group and sex (1907 to 1960).	124

IV. HOUSEHOLDS AND FAMILIES.

TABLES

Number	Title	Page
401.	Total, private, and institutional households by type of household, number of households, and number of persons.	6
402.	Private households by type and size--number of households and number of persons.	7
403.	Institutional households by type and size--number of households and number of persons.	8
404.	Private household population by relatives, sex, and marital status of the head of the household.	9
405.	Private households by type of household and sex, age, and marital status of the head of the household.	10
406.	Private households by sex and nationality of the head of the household and size of household.	13
407.	Private households by occupational major group, age, sex, and employment status of the head of the household.	14
408.	Private households by size of household and size of dwelling.	16
409.	Private households by type of household and size of dwelling.	17

410. Private single family households by socio-economic category, sex and age of the head of the household and size of household. 18

411. Multi-family private households by socio-economic category, sex and age of the head of the household and size of the household. 21

412. Nuclear families by type of household and type of nuclear family. 24

M A L T A

MAIN ENTRIES:

Malta. Central Office of Statistics.
　　1957 census silhouette. Malta, 1959.
　　30 p. 26cm. 　　[Excluded; preliminary data.]

Malta. Central Office of Statistics.
　　Population, housing, and employment census of the Maltese Islands, November 1957: preliminary report. Valetta, 1958.
　　32 p. 33cm. 　　[Excluded; preliminary data.]

Malta. Central Office of Statistics.
　　Census 1957; the Maltese Islands: report on economic activities. [Valetta, 1959?]
　　318 p. 33cm. 　　　　　　　　　　　　　　　　(60-41797)

Malta. Central Office of Statistics.
　　Census 1957; the Maltese Islands: report on population and housing. [Valetta, 1959?]
　　xcviii, 276 p. 33cm. 　　　　　　　　　　　　　(60-23312)

ENUMERATION DATE: November 30, 1957.

ENUMERATION BASIS: De jure.

GLOSSARY

ADMINISTRATIVE, GEOGRAPHICAL, AND POLITICAL UNITS

Level	Vernacular Term
National	Maltese Islands
Intermediate	...
Local	Locality group
	Locality
	Township

　　At the time of the census, Malta was a British colony administered for the Crown by a governor.

259

Locality groups are ten units composed, for census purposes, of localities.

Localities correspond closely to the boundaries of traditional ecclesiastical parishes.* There are 58 localities, of which 52 are composed of a rural/agricultural area and a populated nucleus (township) where most inhabitants, including farm workers, reside.

A township consists of streets fronted by contiguous buildings (or buildings separated by less than 35 yards), and all contiguous areas known to be included in official plans for building development.

TABLE OF DEMOGRAPHIC CONCEPTS

The demographic concepts which appear in the following table correspond to the recommendations of the European Conference of Statisticians, unless otherwise noted. They are defined and discussed in the appendixes and in footnotes. Terms in brackets are not used in the table titles, but they are included here in order to give logical completeness to the various classifications in the tables.

Appendix	English Table Term	Vernacular Table Term
\multicolumn{3}{l}{CULTURAL, EDUCATIONAL, AND PERSONAL CHARACTERISTICS}		
...	Nationality[1]	Same
	Terminal education age[2]	Same
\multicolumn{3}{l}{ECONOMIC CHARACTERISTICS}		
A	Economic activity	Same
	Economically active	Usually gainfully occupied
	Economically inactive	Not usually gainfully occupied

---

*Term not used in table titles.

[1] People are classified by nationality according to their citizenship in one of the following: Malta and its dependencies; United Kingdom; United Kingdom Dominion or Colony; or a specific foreign country.

[2] People who have ceased pursuing full-time education are classified by terminal education age, which is their age (years at last birthday) at the time they last attended school. Full-time education is defined as attendance at school, college, or other educational institution for such time as would leave no reasonable opportunity for regular employment during term time.

TABLE TITLES BY VOLUME

B	Employment status	Same
C	Industrial classification[3]	Same
	Division(s)	Industry
	Major group(s)	Industry
	Group(s)	Industry
D	Occupational classification[4]	Same
	Major group(s)	Main occupational groups
	[Minor group]	...
	Unit group(s)	Individual occupations
...	Socio-economic category[5]	Same
HOUSEHOLD AND FAMILY CHARACTERISTICS		
F	Household	Same

TABLE TITLES BY VOLUME

CENSUS 1957; THE MALTESE ISLANDS: REPORT ON ECONOMIC ACTIVITIES.

TABLES

Number	Title	Page
E.i.	Maltese population by individual occupations, sex, and age ranges.	1
E.ii A.	Maltese persons usually gainfully occupied by main occupational groups and socio-economic groups.	9
E.ii B.	Maltese persons usually gainfully occupied by individual occupations and employment status.	11
E.iii A.	Maltese population showing persons not usually gainfully occupied by type of activity and by sex for localities and locality groups.	36
E.iii B.	Maltese population showing persons usually gainfully occupied by socio-economic groups and by sex for localities and locality groups.	40

---

[3] In the table titles, "industry" designates tabulations of divisions, major groups, or groups. Usage varies.

[4] In the table titles, "selected occupations" designates an abbreviated list of individual occupations.

[5] The classification by socio-economic category groups broad occupational categories into manual/non-manual divisions.

E.iii C. Maltese population showing persons usually gainfully occupied by main occupations and occupational groups and by sex for localities and locality groups. 45

E.iv. Maltese population showing persons usually gainfully occupied by sex and age groups and by marital status in selected occupations. 125

E.v A. Maltese civilians usually gainfully occupied showing, for occupational groups and selected occupations by sex, whether at work, absent from work, or unemployed during the period 28th October to 9th November 1957. 145

E.v B. Part-time workers during the period 28th October to 9th November 1957 showing occupation during that time. 150

E.vi. Civilians having a regular job during the period 28th October to 9th November 1957 by sex and employment status for industries and selected occupations. 153

E.vii. Maltese civilians having a regular job during the period 28th October to 9th November 1957 by sex and industry. 192

E.viii. Maltese civilians aged 14 and over having a regular job during the period 28th October to 9th November 1957 in each group of industries by sex, age groups and marital status. 240

E.ix. Terminal education ages by sex and current age showing the gainfully occupied population by socio-economic occupational groups. 249

E.x. Persons reporting subsidiary occupations showing their main occupations. 254

E.xi. Maltese civilian wage and salary earners having a regular job during the period 28th October to 9th November 1957 by sex, wage/salary group and occupation. 263

E.xii. Maltese civilian wage and salary earners having a regular job during the period 28th October to 9th November 1957 by sex, age-ranges having weekly wage or salary for groups of industries and socio-economic groups. 271

E.xiii. Non-Maltese civilians by country of nationality or citizenship and by sex and socio-economic group. 297

CENSUS 1957; THE MALTESE ISLANDS: REPORT ON POPULATION AND HOUSING.

TABLES

Number	Title	Page
P.I.	Total population by sex, single years of age, and marital status for the Maltese Islands, Malta and Gozo (with Comino).	1
P.II.	Total population by sex, single years of age for ages 0 to 24; quinary age groups, and marital status for locality groups and localities.	13
P.III A.	Density of population (persons per sq. mile) 1921, 1931, 1948, and 1957 for the Maltese Islands, Malta, locality groups and localities.	36
P.III B.	Area and density (persons per sq. mile) showing townships separately for the Maltese Islands, Malta, locality groups and localities.	38
P.IV.	Maltese population aged 14 and over and 20-24 by terminal education ages, and sex for the Maltese Islands, Malta, locality groups and localities.	41
P.V.	Population showing sex and sex-ratios (males per 1000 females) at census dates from 1901 to 1957 for the Maltese Islands, Malta, locality groups and localities.	51
P.VI.	Changes in population, 1842 to 1957, for the Maltese Islands.	57
P.VII.	Total population at successive censuses 1901-1957 by quinary age groups and sex, for the Maltese Islands, Malta and Gozo (with Comino).	58
P.VIII.	Private households (excluding those in Kerrejjas) with 2 or more married couples showing density of occupancy (persons per room) by the joint households, for the Maltese Islands, Malta, Gozo and Comino and localities.	61
P.IX.	Maltese husbands' and wives' ages in combination (couples enumerated on the same schedule), for the Maltese Islands, Malta and Gozo (with Comino).	63
P.X.	Total population by country of nationality or citizenship by sex and four age ranges, for the Maltese Islands.	64

THE NETHERLANDS

MAIN ENTRY AND VOLUME NOTES:[1]

Netherlands (Kingdom, 1815-   ) Centraal Bureau voor de
  Statistiek.
  13^e [i.e., Dertiende] Algemene volkstelling, 31 mei 1960.
[Various places.] 1963-
  v. 29cm.   (NUC67-50140)

Deel 1. Inleiding tot de voornaamste onderwerpen van de telling.
          Bijlage bij deel 1: Begrippen en definities. [An-
          nounced but not yet published.]

Deel 2. Bevolking van de gemeenten en onderdelen van gemeenten.
          1964. 196 p.

Deel 3. Geboortegemeente en periode van vestiging in de huidige
          woongemeente. 1966. 61 p.

Deel 4. Geslacht, leeftijd en burgerlijke staat. 1965. 117 p.

Deel 5. Huishoudens, gezinnen en woningen. A. Algemene
          inleiding. [Announced but not yet published.]
          B. Voornaamste cijfers per gemeente. 1964. 97 p.

Deel 6. Bestaande huwelijken en vruchtbaarheid van deze
          huwelijken. 1969. 57 p.

Deel 7. Kerkelijke gezindte. A. Algemene inleiding. 1968.
          106 p.   B. Voornaamste cijfers per gemeente. 1963.
          93 p.

Deel 8. Genoten onderwijs en opleidingsniveau. A. Algemene
          inleiding. 1969. 87 p.   B. Voornaamste cijfers
          per gemeente. 1964. 117 p.

Deel 9. Academish gevormden. 1964. 93 p.

Deel 10. Beroepsbevolking. A. Algemene inleiding. 1967. 151 p.
          B. Voornaamste cijfers per gemeente. 1964. 163 p.
          C. Vergelijking van de uitkomsten van de beroepstel-
          lingen 1849-1960. 1966. 52 p.

Deel 11. Buiten de woongemeente werkenden. 1965. 105 p.

---

[1] The volume and table titles of the Netherlands census are
given in both Dutch and English. Each volume contains an English
summary of the introductory material for that volume.

MAIN ENTRY AND VOLUME NOTES

Deel 12. Bevolking in inrichtingen en tehuizen. 1966. 79 p.
Deel 13. Varende en rijdende bevolking. 1965. 40 p.
Deel 14. Voornaamste kengetallen per gemeente. 1966. 99 p.

ENGLISH TRANSLATION OF CENSUS AND VOLUME TITLES:
Thirteenth general population census, May 31, 1960.

1. Introduction to the main items of the census. Annex: Concepts and definitions. [Announced but not yet published.]
2. Population of the municipalities and the territorial subdivisions of the municipalities.
3. Municipality of birth and period of settlement in the present municipality of residence.
4. Sex, age, and marital status.
5. Households, families, and dwellings. A. General information. [Announced but not yet published.] B. Principal data for each municipality.
6. Existing marriages and marital fertility.
7. Religion. A. General introduction. B. Principal data for each municipality.
8. Type of education received and overall level of education. A. General instruction. B. Principal data for each municipality.
9. University graduates.
10. Economically active population. A. General introduction. B. Principal data for each municipality. C. Comparison of the general censuses of population 1849-1960 as far as the economically active population is concerned.
11. Economically active population working outside their municipality of residence.
12. Population living in institutions.
13. Population permanently living on inland vessels and in caravans.
14. Principal key figures for each municipality.

ENUMERATION DATE: May 31, 1960.

ENUMERATION BASIS: De jure.

THE NETHERLANDS

GLOSSARY

ADMINISTRATIVE, GEOGRAPHICAL, AND POLITICAL UNITS

Level	English Translation	Vernacular Term
National	The Netherlands	Koninkrijk der Nederlanden
Intermediate	Provinces	Provincie
Local	Municipality (commune)	Gemeente
Population cluster ("locality") concepts:[1]		
	Places/Population concentrations	Plaatsen
	Scattered houses	Verspreide huizen
	Agglomerations	Agglomeraties
Other concept:		
	Polder boards	Waterschappen

The Kingdom of the Netherlands is divided into 11 provinces which are divided into a total of 994 municipalities, and these in turn into about 8,000 territorial subdivisions. Within each municipality, a distinction is made between subdivisions coinciding with built-up areas (places or population clusters) and subdivisions comprising areas of scattered houses. Places or population concentrations are territorial groups of houses forming more or less continuous built-up areas and having locally recognized place-names. The population living outside these clusters is classified as living in scattered houses.

Agglomerations are two or more originally separated places which have grown together. They may extend across two or more adjoining municipalities.

Polder boards are the bodies which oversee Ijsselmeer polders. Polders are units of land which have been reclaimed from the Ijsselmeer (formerly called the Zuiderzee).

TABLE OF DEMOGRAPHIC CONCEPTS

The demographic concepts which appear in the following table correspond to the recommendations of the European Conference of Statisticians, unless otherwise noted. They are defined and discussed in the appendixes and in footnotes. Terms in parentheses

---

[1] For a discussion of this concept, see European Population Censuses: The 1960 Series, U.N. Doc.: ST/CES/3 (1964), p. 108.

GLOSSARY

are alternative forms which appear in the table titles. Terms in brackets are not used in the table titles, but they are included here in order to give logical completeness to the various classifications in the table.

Appendix	English Table Term	Vernacular Table Term
	CULTURAL, EDUCATIONAL, AND PERSONAL CHARACTERISTICS	
...	Education[1]	Onderwijs
	Religion[2]	Kerkelijke gezindte
	ECONOMIC CHARACTERISTICS	
A	Economic activity	Beroepsarbeid
	Economically active[3]	Beroepsbevolking
	Economically inactive	Bevolking zonder beroep

---

[1] All persons except those with a university education were asked to give information about all kinds of schools attended and all certificates held. Persons with a university education were asked to give only the most advanced examination passed at the university longest attended. All persons 14 years of age and older not attending school were classified into each of three categories of educational qualifications, or <u>types of education</u>: (1) General education (primary school, secondary modern school, secondary grammar school or junior seminary, and university); Lower, Upper-lower, Medium, or Higher (Bachelor's and Master's Degrees); (2) Other full-time education, mostly vocational in character: Lower, Medium, or Semi-higher; (3) Certificates and diplomas not falling under (1) or (2): Lower, Medium, or Semi-higher. Only the highest educational level was considered, and by combining for each individual the level of education attained in each of the three categories, an overall level of education was established. The overall level of education fell into one of five categories: Lower, Upper-lower, Medium, Semi-higher, or Higher.

[2] Data on religious affiliation derive from a question about the denomination to which the individual belonged. The respondent could enter a different religion from the official one in which he was baptized, or he could claim "no religious affiliation." These results accordingly differ from those supplied by the various major religious denominations, which include the Roman Catholic Church, the Netherlands Reformed Church, Calvinist churches, Baptist churches, the Evangelical Lutheran Church, the Remonstrant Church, and Judaism. "Other" denominations and sects include groups such as the Salvation Army and Jehovah's Witnesses.

[3] Persons employed and unemployed at the time of the census having an occupation in which they are generally engaged for at least 15 hours per week.

THE NETHERLANDS

B	Employment status (Industrial status)		Positie in het bedrijf

C	Industrial classification
	  Division(s)
	  Major group(s)
	  Group(s)
	  Subgroup(s)
			Bedrijf
			  Bedrijfstak
			  Bedrijfklasse
			  Bedrijfgroep
			  Bedrijfssubgroep

D	Occupational classification
	  [Major group]
	  Minor group(s)
	  Unit group(s)
	  Unit subgroup(s) (Occupation)
			Beroep
			
			  ...
			  Beroepsklasse
			  Beroepsgroep
			  Beroepsnaam

E	Socio-economic category (Social status, socio-economic group, socio-economic status)		Sociale beroepsgroep

HOUSEHOLD AND FAMILY CHARACTERISTICS

F	Households
	  One-person households
	  Multi-person households
		Huishoudens
		  Alleenstaaden
		  Huishoudens

	Institutions	Inrichtingen
	Family nuclei	Gezinnen

TABLE TITLES BY VOLUME

VOLUME 2. POPULATION OF THE MUNICIPALITIES AND OF THE TERRITORIAL SUBDIVISIONS OF THE MUNICIPALITIES.

TABLES

Number	Title	Page
1.	Total population, area and population density of the municipalities and number of inhabitants of the largest place, of the other places (according to size-groups) and in scattered houses. By province in alphabetical order.	21
2.	Population of municipalities and places. In alphabetical order.	69
3.	Population of places or "agglomerations" (two or more places forming a continuous built-up area). By province according to decreasing size of population.	117
4.	Area and population density in population concentrations having 2,000 inhabitants or more.	169

TABLE TITLES BY VOLUME

VOLUME 3. MUNICIPALITY OF BIRTH AND PERIOD OF SETTLEMENT IN THE PRESENT MUNICIPALITY OF RESIDENCE.

TABLES

Number	Title	Page
1.	Population per province of residence according to place of birth.	10
2.	Population per province of residence according to period of settlement in the present municipality of residence.	11
3.	Total population and those locally born or having always lived in their present municipality of residence according to age and sex.	12
4.	Total population and those locally born according to the degree of urbanization (1960) of the present municipality of residence.	15
5.	"Friezen"[1] and "Brabanders"[2] living outside their province of birth according to the degree of urbanization of the present municipality of residence.	17

VOLUME 4. SEX, AGE, AND MARITAL STATUS.[3]

TABLE

Title	Page
Population by sex, age, and marital status per municipality, May 31, 1960.	42

---

[1] Persons born in the province of Friesland.

[2] Persons born in the province of Noord-Brabant.

[3] Age was calculated from the year of birth entered on the questionnaire. Persons born in 1959 and 1960 were regarded as 0 years old, in 1958 as 1 year old, etc. Actually those born from January to May 1959 were all 1 year old at the time of the census. This method, therefore, overstates the number in the age 0 category and understates the number in the oldest age group. The method had little effect on the remaining age groups, however, and hence the difference from the actual age distribution is very small. For marital status, the term "married" includes those judicially separated.

VOLUME 5. HOUSEHOLDS, FAMILIES,[1] AND DWELLINGS. B. PRINCIPAL DATA FOR EACH MUNICIPALITY.

TABLES

Number	Title	Page
1.	Population, multi-person households (definition 1960) and one-person households (definition 1960) per municipality.	40
2.	Multi-person households (definition 1947), one-person households (definition 1947), family nuclei, and housing stock,[2] per municipality.	60

VOLUME 6. EXISTING MARRIAGES AND MARITAL FERTILITY.

TABLES

Number	Title	Page
1.	Existing first and subsequent marriages according to year of marriage and age of husband resp.[3] wife at marriage.	22
2.	Existing first and subsequent marriages according to period in which marriage was contracted and mutual age of spouses at marriage.	23
3.	Existing first and subsequent marriages according to mutual age of spouses on May 31st, 1960.	24
4.	Existing first and subsequent marriages according to period in which marriage was contracted and difference in age of spouses.	25

---

[1] The definition of <u>household</u> used here, which is based on the one recommended by the European Conference of Statisticians, differs from that previously used in the Netherlands. Hence, to insure comparability, data for both definitions are presented. The old definition is "definition 1947"; the new, "definition 1960." According to the former every family nucleus constituted a separate household. According to the latter, however, the household is considered to be a residential unit which may contain more than one family.

[2] "Housing stock" is a term for available buildings, inhabited and vacant, distinguished into conventional dwellings, mobile housing units, or other housing units.

[3] Ages of husband and wife respectively.

5.	Existing first and subsequent marriages according to period in which marriage was contracted and mutual religion of spouses.	26
6.	Existing first marriages according to period in which marriage was contracted, mutual denomination, and difference in age of spouses.	28
7.	Existing first marriages according to period in which marriage was contracted, mutual denomination of spouses, and socio-economic status of husband.	31
8.	Existing first and subsequent marriages according to year (period) in which marriage was contracted and number of live-born children per marriage.	34
9.	Existing first and subsequent marriages according to age of wife at marriage and number of live-born children per marriage.	35
10.	Existing first and subsequent marriages according to denomination of wife and number of live-born children per marriage.	36
11.	Existing first marriages according to period in which marriage was contracted, age of wife at marriage, and number of live-born children.	37
12.	Existing first marriages according to socio-economic status of husband, period in which marriage was contracted, age of wife at marriage, denomination of wife, and number of live-born children.	38
12.A.	Existing first marriages of university graduated men according to period in which marriage was contracted, age of wife at marriage, denomination of wife, and number of live-born children.	54
13.	Existing first marriages according to period in which marriage was contracted, mutual denomination of spouses, and number of live-born children.	55

VOLUME 7. RELIGION. A. GENERAL INTRODUCTION.

TABLES

Number	Title	Page
1.a.	Population by religious denomination, sex, and age.	41
1.b.	Age-distribution and sex-proportion by religious denomination.	42
1.c.	Age-distribution by religious denomination, by sex.	43
1.d.	Distribution by religious denomination, by age-group, by sex.	43

1.e.	Sex-proportion by age-group, by religious denomination.	44
2.a.	Population (some smaller religious denominations) by sex and age.	44
2.b.	Age-distribution and sex-proportion for some smaller religious denominations.	45
3.a.	Population by religious denomination, sex, age, and civil status.[1]	46
3.b.	Unmarried persons of 15 years and older by religious denomination, age, and sex.	50
3.c.	Married persons by religious denomination, age, and sex.	51
3.d.	Judicially separated persons by religious denomination, age, and sex.	51
3.e.	Widowed persons by religious denomination, age, and sex.	52
3.f.	Divorced persons by religious denomination, age, and sex.	53
4.a.	Economically active population by religious denomination, social status, and sex.	54
4.b.	Economically active population. Distribution according to social status, by religious denomination, by sex.	56
5.a.	Heads of households (definition 1947) by religious denomination, social status, and sex.	58
5.b.	Heads of households (definition 1947). Distribution according to social status, by religious denomination, by sex.	60
6.a.	Economically active male population by religious denomination, social status, and age.	62
6.b.	Economically active male population. Distribution according to social status, by religious denomination, by age-group.	62
6.c.	Economically active male population. Distribution according to religious denomination, by age-group, by social status.	64
6.d.	Economically active male population. Distribution according to age-group, by religious denomination, by social status.	64

---

[1] Marital status.

7.a.	Economically active population by religious denomination, educational status,[2] and sex.	66
7.b.	Economically active male population. Distribution according to religious denomination, by educational status,[2] by age-group.	73
7.c.	Economically active female population. Distribution according to religious denomination, by educational status,[2] by age-group.	74
7.d.	Economically active male population. Distribution according to educational status,[2] by religious denomination, by age-group.	75
7.e.	Economically active female population. Distribution according to educational status,[2] by religious denomination, by age-group.	76
7.f.	Economically active male population. Distribution according to educational status,[2] by age-group, by religious denomination.	77
8.a.	Economically active male population by religious denomination, social status, educational status,[2] and age.	78
8.b.	Economically active male population. Distribution according to social status, by educational status,[2] by age-group, by religious denomination.	84
8.c.	Economically active male population. Distribution according to social status, by age-group, by religious denomination.	88
8.d.	Economically active male population. Age-distribution by social status, by educational status,[2] by religious denomination.	92
8.e.	Economically active male population. Distribution according to religious denomination, by social status, by educational status,[2] by age-group.	96
9.a.	Persons of 14 years and older without occupation, not attending school, by religious denomination, general education, sex, and age.	100
9.b.	Persons of 14 years and older without occupation, not attending school. Age-distribution by general education, by religious denomination, by sex.	101
9.c.	Persons of 14 years and older without occupation, not attending school. Distribution according to	

---

[2] Types of education. (See Table of Demographic Concepts: "Education.")

general education, by age-group, by religious denomination, by sex. 102

VOLUME 7. RELIGION. B. PRINCIPAL DATA FOR EACH MUNICIPALITY.

TABLES

Number	Title	Page
1.	Population by religion, by municipality.	36
2.	Further breakdown of the group "other" in Table 1, for larger municipalities and groups of smaller municipalities.	82

VOLUME 8. TYPE OF EDUCATION RECEIVED AND OVERALL LEVEL OF EDUCATION. A. GENERAL INTRODUCTION.

TABLES

Number	Title	Page
1.a.	Population of 14 years and over, not attending school, with and without occupation, according to general education, sex, and age.	16
1.b.	Population of 14 years and over. Age-distribution by type of general education, by sex.	17
1.c.	Population of 14 years and over. Distribution according to general education, by age-group, by sex.	18
2.a.	Economically active population according to full-time vocational education, sex, and age.	19
2.b.	Economically active population. Age-distribution by type of full-time vocational education, by sex.	20
2.c.	Economically active population. Distribution according to full-time vocational education, by age-group, by sex.	21
3.a.	Economically active population according to certificates and diplomas, overall level of education, and sex.	22
3.b.	Economically active persons with certificates or diplomas. Distribution according to type of certificate by education by sex, and according to type of education by certificate by sex.	24
4.	Persons of 14 years and over, not attending school and without occupation, according to certificate, sex, and age.	25

5.a.	Economically active population according to overall level of education, age, and sex.	26
5.b.	Economically active population. Distribution according to overall level of education, by age-group, by sex.	28
5.c.	Economically active population. Age-distribution by overall level of education, by sex.	29
6.a.	Economically active population according to overall level of education, social status, and sex.	30
6.b.	Economically active population. Distribution according to overall level of education, by social status, by sex.	32
6.c.	Economically active population. Distribution according to social status, by overall level of education, by sex.	33
7.a.	Economically active population according to overall level of education, social status, age, and sex.	34
7.b.	Economically active population. Distribution according to overall level of education, by age-group, by sex, by social status.	47
7.c.	Economically active population. Age-distribution by overall level of education, by sex, by social status.	52
7.d.	Economically active population. Distribution according to social status, by overall level of education, by sex, by age-group.	57
8.a.	Economically active population according to overall level of education, occupation, and sex.	62
8.b.	Economically active population. Distribution according to occupation, by overall level of education, by sex.	70
8.c.	Economically active population. Distribution according to overall level of education, by occupation, by sex.	72

VOLUME 8. TYPE OF EDUCATION RECEIVED AND OVERALL LEVEL OF EDUCATION. B. PRINCIPAL DATA FOR EACH MUNICIPALITY.

TABLE

Title   Page

Table with figures for each municipality. Male and female population of 14 years of age and over: with an occupation, according to type of education received and overall level of education; without an

THE NETHERLANDS

occupation, not attending school, according to type of general education received; still attending school. 38

VOLUME 9. UNIVERSITY GRADUATES.[1]

TABLES

Number	Title	Page
1.	University-trained persons with and without an occupation by department of study, sex, and examination passed.	46
2.	University graduates by department of study, sex, and place of study.	48
3.	University graduates by department of study, sex, and age.	50
4.	University graduates by department of study, sex, age, and religious denomination.	54
5.	University graduates with an occupation by department of study, sex, and age.	68
6.	University graduates with an occupation by department of study, sex, and industrial status.	72
7.	University graduates employed in the public service (excluding armed forces on compulsory military service) by department of study and sex.	74
8.	University graduates (employees only and excluding armed forces on compulsory military service) by department of study, sex, engagement,[2] and annual income.	76
9.	University graduates with an occupation by kind of industrial activity[3] and by department of study.	80

---

[1] All university graduates were asked whether they attended a Dutch or foreign university, in which university and department they studied, their major subject, and the highest examination passed. Included in this volume are all persons who passed a final examination. The departments of study are: medical sciences; social sciences; natural sciences; theology; and arts and letters.

[2] A modified industrial status adapted to the special occupations of university graduates, classifying them as persons in authority, persons teaching, or others working as employees.

[3] Industrial group.

| 10. | University graduates by kind of occupation and by department of study. | 88 |

VOLUME 10. ECONOMICALLY ACTIVE POPULATION. A. GENERAL INTRODUCTION.

TABLES

Number	Title	Page
1.	Economically active and inactive population by age, sex, and marital status.	56
2.	Economically active population by major group of industry, industrial status, and sex, 1947 and 1960.	58
3.	Economically active population by subgroup of industry, and sex, 1947 and 1960.	72
4.	Economically active population by major group of industry, age, and sex (5% sample).	82
5.	Inactive population by nationality and economically active population by nationality, industrial status, division of industry, and sex.	85
6.	Economically active population by minor group of occupation, industrial status, and sex.	86
7.	Economically active population by unit group of occupation, industrial status, and sex; employers and own account workers, respectively; employees by age and sex.	90
8.	Economically active population by occupation and sex (professions and employees only), 1947 and 1960.	110
9.	Economically active population by socio-economic group, place in household, age, and sex.	114
10.	Heads of household by socio-economic group, sex, and number of dependent children or other dependents residing with the family.	119
11.	Heads of families by socio-economic group, sex, and number of economically active children residing with the family.	121
12.	Male heads of families by socio-economic group in relation to industrial status, full- or part-time employment of economically active wife, and presence of independent and/or dependent children residing with the family.	122
13.	Economically active male population by socio-economic group, educational level, and age.	124

14. Government workers employed by State, province, municipality, polder board, or public industrial corporation, by industrial status, industry,[1] and sex.   125

15. Government workers employed by State, province, municipality, polder board, or public industrial corporation, income-class, religious denomination, age, and sex.   127

16. Female employers and own account workers by industry,[2] marital status, place in household or family.   129

17. Female employees by minor group of occupation, marital status, place in household or family, and full- or part-time employment.   130

18. Economically active female population by completed general education or vocational training; married economically active females, husband present, by industrial status and completed general education or vocational training, and age.   134

19. Married economically active female population by age, 1947 and 1960.   135

20. Married economically active female population, husband present, by major group of industry and industrial status, 1947 and 1960.   136

21. Married economically active female population, husband present, by occupation, 1947 and 1960.   138

22. Married economically active female population, by socio-economic group, age and number of children under 6 years of age residing with family, and full- or part-time employment.   140

VOLUME 10. ECONOMICALLY ACTIVE POPULATION. B. PRINCIPAL DATA FOR EACH MUNICIPALITY.[3]

TABLES

Number	Title	Page
1.	Economically active population by kind of economic activity[4] and sex, by municipality and province.	42

---

[1] Major group and group of industry.

[2] An abbreviated list of divisions and major group levels of industry.

[3] This volume provides data concerning only the <u>resident</u> population of each municipality, including residents working outside the municipality.

[4] An abbreviated industrial classification comprising four groups: agriculture, industry, business, and services.

TABLE TITLES BY VOLUME

2. Economically active population by industrial status and sex, by municipality and province.   84

3. Male heads of household by socio-economic group, by municipality and province.   126

VOLUME 10. ECONOMICALLY ACTIVE POPULATION. C. COMPARISON OF THE RESULTS OF THE GENERAL CENSUSES OF POPULATION 1849-1960 WITH RESPECT TO THE ECONOMICALLY ACTIVE POPULATION.

TABLES

Number	Title	Page
1.	Economically active population by kind of economic activity (major and subgroups of industry) and sex, 1899-1960.	28
2.	Economically active population by sex and industrial status in some divisions of industry.	40
3.	Employees (salaried employees and wage-earners) and professional workers by occupation, 1920-1960.	42

VOLUME 11. ECONOMICALLY ACTIVE POPULATION WORKING OUTSIDE THEIR MUNICIPALITY OF RESIDENCE.[1]

TABLES

Number	Title	Page
1.	Resident economically active population, number of outgoing commuters, others working outside the municipality of residence, number of incoming commuters, surplus incoming over outgoing commuters, and economically active population employed locally, by municipality.	51

---

[1] Persons employed outside the municipality of residence were divided into two categories: (1) commuters proper, persons traveling daily to only one other municipality to work, and (2) others employed outside their municipality of residence, i.e., persons not commuting daily and/or having no fixed municipality of work. Any commuter proper can be regarded as both an outgoing commuter and an incoming commuter, depending on whether he is classified with respect to the municipality in which he resides or the municipality in which he works. The chief criterion for a commuter proper is the daily crossing of the municipal boundary. The means of transportation considered were: bicycle, bicycle and auxiliary engine, bus, train, and automobile.

THE NETHERLANDS

2. Outgoing and incoming commuters according to direction and means of transportation; outgoing commuters according to home-to-work distance (in minutes) and period in which they moved into the present municipality of residence, as well as some additional data concerning the outgoing male commuters, by municipality.   66

3. Outgoing and incoming commuters according to branch of economic activity,[2] by municipalities with 100,000 or more inhabitants.   96

VOLUME 12. POPULATION LIVING IN INSTITUTIONS.[3]

TABLES

Number	Title	Page
1.	Population in institutions, by municipality.	34
2.	Amboynese living in camps, etc., by municipality.	64
3.	Inmates in institutions by (main) purpose of the institution, by municipality.	66

VOLUME 13. POPULATION PERMANENTLY LIVING ON INLAND VESSELS AND IN CARAVANS.[4]

TABLE

Title	Page
Table of the population entered in a municipal register or in the Central Population Register, by municipality of registration and sex.	29

---

[2] Industrial major group.

[3] Persons living permanently in institutions were enumerated separately with special questionnaires. The definition of **institution** is: all places accommodating persons who must submit to the regulations of the place, for a special purpose under general guidance. The institutional population includes all inhabitants of all dwellings and other housing units situated on the grounds. Persons living in institutions are divided into two groups: (1) inmates and (2) staff and management and their household members. Amboynese are persons ejected from Indonesia when it was freed from Dutch rule. The purposes of the institutions enumerated in Table 3 are: institutions for the elderly; hospitals; mental hospitals; institutions for the physically handicapped; orphanages; welfare institutions; boarding schools; cloisters; military and police barracks; prisons; and others.

[4] The population living permanently on inland vessels and in caravans was enumerated separately. The mobility of these

TABLE TITLES BY VOLUME

VOLUME 14. PRINCIPAL KEY FIGURES FOR EACH MUNICIPALITY.[1]

TABLES

Number	Title	Page
1.	Population, households, families, and housing.	20
2.	Economically active population and commuting.	54

---

population groups necessitated a number of qualifying definitions. All figures relate to the population on inland vessels and in caravans who belonged to the Netherlands population, i.e., only those who were registered in one of the municipal population registers or in the Central Population Register. The population on inland vessels includes independent shipmasters, wage-earning shipmasters, crew members, and persons (living on board) belonging to the households of the former. The population living in caravans includes traditional caravan populations (gypsies, etc.), persons employed by a circus or a travelling fair, and persons employed by construction and related industries (particularly road construction). The table for this volume gives the number of persons in caravans and on inland vessels by sex and by municipality in which they are registered.

[1]This volume summarizes some key statistics concerning the subjects in the table titles. These statistics are of special importance for comparative regional studies, where statistical indicators of differences and similarities among municipalities are often needed.

NORTHERN IRELAND

MAIN ENTRY AND VOLUME NOTES:

Northern Ireland. General Register Office.
   Census of population, 1961. Belfast, 1962-1965.
   10 v. 33cm. (65-40135)

   General report. 1965. lxi, 84 p.
   Fertility report. 1965. x, 50 p.
   Topographical index. 1962. 189 p. [Excluded; no tables.]
   Belfast County Borough. 1963. xxxiv, 64 p.
   County of Antrim. 1964. xxxii, 75 p.
   County of Armagh. 1964. xxxiv, 56 p.
   County of Down. 1964. xxxv, 77 p.
   County of Fermanagh. 1964. xxxii, 49 p.
   County and County Borough of Londonderry. 1964. xxxvi, 92 p.
   County of Tyrone. 1964. xxxii, 65 p.

ENUMERATION DATE: April 23, 1961.

ENUMERATION BASIS: De facto.

GLOSSARY

ADMINISTRATIVE, GEOGRAPHICAL, AND POLITICAL UNITS

Level	Vernacular Term
National	Northern Ireland
Intermediate	...
Local	County/County borough
	Municipal borough
	District
	Urban
	Rural
	Urban and rural aggregates

GLOSSARY

Other concepts:

> Parliamentary constituency
> Town
> Ward

Northern Ireland is divided into 6 <u>counties</u> and 2 <u>county boroughs</u>: Belfast and Londonderry.

The 6 counties are divided into a total of 65 administrative areas: 9 <u>municipal boroughs</u>, 25 <u>urban districts</u>, and 31 <u>rural districts</u>.

<u>Boroughs</u> are incorporated municipalities (unlike towns), and <u>urban</u> and <u>rural districts</u> are the divisions of the overall administrative areas of the counties. The term <u>town</u> refers to towns, villages, or housing estates consisting of separate and identifiable clusters of 50 or more dwellings without legally defined boundaries; towns forming continuous housing developments in administrative areas are excluded.

The 2 <u>county boroughs</u> are large municipalities, each with a single council responsible for all services within its area.

<u>Urban aggregates</u> are composed of the 2 county boroughs, plus the municipal boroughs and urban districts of the counties. <u>Rural aggregates</u> are the combined rural districts of the counties.

There are 48 territorial <u>parliamentary constituencies</u> for elections to the Northern Ireland Parliament, each returning one member. In addition, 4 members are returned by the constituency of the Queen's University of Belfast. There are 12 constituencies for elections to the United Kingdom Parliament.

The 2 county boroughs, together with 6 municipal boroughs and 11 urban districts, are divided into <u>wards</u>, the units for elections to the respective local councils. For the remaining 3 municipal boroughs and 14 urban districts, the entire borough or urban district is the unit for council elections. In the rural districts, the district electoral divisions are the units for local council elections.

NORTHERN IRELAND

TABLE OF DEMOGRAPHIC CONCEPTS

The demographic concepts which appear in the following table correspond to the recommendations of the European Conference of Statisticians, unless otherwise noted. They are defined and discussed in the appendixes and in footnotes. Terms in brackets are not used in the table titles, but they are included here in order to give logical completeness to the various classifications in the table.

Appendix	English Table Term	United Kingdom Terminology
CULTURAL, EDUCATIONAL, AND PERSONAL CHARACTERISTICS		
...	Religion[1]	Same
...		Terminal education age[2]
ECONOMIC CHARACTERISTICS		
A	Economic activity	Same
	Economically active	Occupied
	Economically inactive	Same
B	Employment status	Industrial status
C	Industrial classification[3]	Same
	Division(s)	Orders
	Major group(s)	Orders
	Group(s)	Minimum list headings
	Subgroup(s)	Sub-divisions
D	Occupational classification[4]	Same
	Major group(s)	Orders
	Minor group(s)	Unit groups
	[Unit group][5]	...
...	...	Social class[6]
E	Socio-economic category[7]	Same
HOUSEHOLD AND FAMILY CHARACTERISTICS		
F	Household[8]	Same
	Private household[9]	Same
	Family[10]	Same

---

[1][All footnotes for the Table are on the following pages.]

GLOSSARY

[1] Religion is divided into the following five categories: Roman Catholic, Presbyterian, Church of Ireland, Methodist, and Other and "Not stated" denominations.

[2] Terminal education age is the age at which all persons 15 years of age or older, who completed education at school, college, or university, ceased receiving full-time instruction.

[3] In the table titles, "industry" designates tabulations of any and all levels of industrial classification. The U.K. system of industrial classification corresponds to the ISIC 1958 as follows: "Orders" are "Divisions" and "Major groups"; "Minimum list headings" are "Groups"; and "Sub-divisions" are "Subgroups."

[4] In the table titles, "occupation" designates tabulations of the ISCO 1958 major and minor groups together, or of the major groups alone. In the U.K. system of occupational classification, "Orders" are "Major groups," and "Unit groups" are "Minor groups." For a detailed discussion of the comparability of the U.K. occupational classification system to that of the ISCO 1958, see Great Britain, <u>Classification of Occupations,</u> 1960, p. vi.

[5] U.K. use of the term "Unit group" is not to be confused with the ISCO 1958 definition of the term.

[6] The classification by social class groups the employed population into the following occupational categories:
    a. Professional, etc., occupations (all non-manual)
    b. Intermediate occupations:
        1. Manual
        2. Non-manual
        3. Agricultural
    c. Skilled occupations:
        1. Manual
        2. Non-manual
        3. Agricultural
    d. Partly skilled occupations:
        1. Manual
        2. Non-manual
        3. Agricultural
    e. Unskilled occupations (all manual).

[7] The classification by socio-economic category is similar, but not identical, to that recommended by the European Conference of Statisticians.

[8] A household is one person living alone, or a group of persons living together who normally take meals together, whether in a dwelling or part of one, or in a hotel, boarding house, hospital,

service establishment, or other premises providing some form of residential accommodation.

[9] A private household is a household occupying all or part of a house, flat, farmhouse, or other type of dwelling. A person (or persons) living in the same dwelling as, but not boarding with, the household is treated as a separate household. If such a person is usually provided with at least one meal per day by the household, he is regarded as part of it.

[10] Tabulations of family characteristics other than size are not given.

TABLE TITLES BY VOLUME

GENERAL REPORT.

TABLES

Number	Title	Page
1.	Area, buildings for habitation and population, 1961.[1]	1
2.	Population, 1821-1961.	1
3.	Population, 1951 and 1961, and intercensal changes.[1]	1
4.	Area, population, buildings for habitation, private households and valuation.[2]	2
5.	Private dwellings: inhabited dwellings by building type, tenure and rooms.[3]	6
6.	Private households: size, rooms occupied, and density of room occupation.[3]	10
7.	Private households: availability of household arrangements by tenure and type of accommodation.[3]	12
8.	Private households: households and persons therein by tenure of accommodation, rooms occupied and socio-economic group of head of household.	14

[1] Northern Ireland, counties, and county boroughs.

[2] Northern Ireland, administrative urban and rural aggregates, counties, county boroughs, administrative areas of counties, towns with 1,000 or more population.

[3] Northern Ireland, administrative urban and rural aggregates, counties, and county boroughs.

TABLE TITLES BY VOLUME

9.	Inhabited buildings, etc., other than private dwellings: class, number, total population and (for institutions) number of inmates.[1]	15
10.	Population, Parliamentary Electors and Members of Parliament.[4]	17
11.	Adjustment of enumerated (de facto) population to obtain resident (de jure) population.[5]	19
12.	Ages by single years, sex and marital condition.[6]	20
13.	Ages by quinquennial groups, religion, sex and marital condition.	23
14.	Married couples, enumerated together, by quinquennial age groups of husbands and wives.	25
15.	Birthplaces.	26
16.	Birthplaces: persons born outside Northern Ireland by marital condition and five age groups.[1]	27
17.	Birthplaces: persons born outside Northern Ireland by religion and place of birth.	28
18.	Nationality: persons born outside the British Isles by nationality and usual address a year ago.	28
19.	Religions.[1]	29
20.	Religions: population under 22 years by individual years and 20 years and over by quinquennial groups.	30
21.	Education: terminal education ages of persons 15 years and over by sex and age group.[1]	31
22.	Education: terminal education ages of persons 15 years and over by sex, age group and occupation order.	34
23.	Science and technology: persons with scientific or technological qualifications[7] by subject and type of qualification.	36

---

[4] Parliamentary constituencies.

[5] Counties and county boroughs.

[6] Northern Ireland and administrative urban and rural aggregates.

[7] The question on scientific and technological qualifications was the same as that asked in the censuses for other parts of the United Kingdom. Information was obtained on the <u>type of qualification</u> (university degree, graduate or corporate membership of a professional association, etc.) for persons qualified

24.	Science and technology: persons with scientific or technological qualifications[7] by sex, subject of qualification and age (and by marital condition for occupied females).	37
25.	Science and technology: persons with scientific or technological qualifications[7] by subject and type of qualification and occupation.	38
26.	Science and technology: persons with scientific or technological qualifications[7] by subject and type of qualification and industry.	40
27.	Occupations: population aged 15 and over by occupation and industrial status.	42
28.	Occupations: occupied population aged 15 and over by occupation, marital condition and age group.	48
29.	Occupations: part-time workers and family workers by occupation order, and retired persons by former occupation order.	62
30.	Occupations: population aged 15 and over by socio-economic group, social class and age group.	64
31.	Industries: working population aged 15 and over by industry (excluding persons out of work).	65
32.	Industries: working population aged 15 and over by industry, marital condition and age group (excluding persons out of work).	69
33.	Industries: working population aged 15 and over born outside Northern Ireland by industry and place of birth.	81

FERTILITY REPORT.[1]

TABLES

Number	Title	Page
1.	All married, widowed and divorced women by age at census, duration of marriage and size of family.	2

in science and technology, excluding medicine, dentistry, pharmacy, optics, veterinary science, architecture, economics, geography, and the social sciences, and on the subject of qualification (biology, chemistry, etc.). Unless otherwise noted, the area for which statistics are given is Northern Ireland.

[1]Data in this volume derive from questions on the number of children born alive in marriage and whether any of these children were born in the year before the census. These questions were

TABLE TITLES BY VOLUME

2. Women married once only who married before age 45 plus remarried, widowed and divorced women whose first or only marriage lasted to age 45 by age at census, duration of marriage and size of family.   7

3. Women married once only who married before age 45 plus remarried, widowed and divorced women whose first or only marriage lasted to age 45: such women who had a child in year ended 23 April 1961, by age at census, duration of marriage, size of family and fertility rate.   12

4. All married, widowed and divorced women by age at marriage, duration of marriage and size of family.   15

5. Women married once only who married before age 45 plus remarried, widowed and divorced women whose first or only marriage lasted to age 45 by age at marriage, duration of marriage and size of family.   18

6. Women married once only who married before age 45 plus remarried, widowed and divorced women whose first or only marriage lasted to age 45: such women who had a child in year ended 23 April 1961, by age at marriage, duration of marriage, size of family and fertility rate.   21

7. Women married once only who married before age 45 and who were aged 45 and over at census day and enumerated with their husbands by socio-economic group of husband, religion, age at marriage, duration of marriage and size of family.   24

---

put to all married, widowed, and divorced women. <u>Duration of marriage</u> of married women is computed as the difference in years and months between the date of census and date of first or only marriage; of widowed and divorced women, the difference between the dates of commencement and termination of first or only marriage. Duration is [then] taken as the number of completed years. <u>Age at marriage</u> of married women is age at census in whole years less duration of marriage in years. For widowed and divorced women, age at marriage is the difference in years and months between date of first or only marriage and census date; the residual months are dropped and the result subtracted from the census age in whole years. <u>Size of family</u> is the total number of children born alive to a woman in marriage, including any that have died. <u>Fertility rate</u> in Tables 3 and 6 is the ratio of the number of married women in each census age by marriage duration group who had a live-born child or children during the year ending 23 April 1961, per 1,000 women in the corresponding group in Table 2.

8. Women married once only who married before age 45 and who were aged 45 and over at census day and enumerated with their husbands by age of wife and of husband at marriage, duration of marriage, and mean family size. 41

9. Women married once only who married before age 45 plus remarried, widowed and divorced women whose first or only marriage lasted to age 45: such women aged 45 and over at census day and whose duration of marriage was 25 years and over by age at marriage, year of marriage, religion and mean family size. 48

BELFAST COUNTY BOROUGH . . . COUNTY OF TYRONE.
[7 volumes; see listing on p. 282 above.]

Structure of the County and County Borough Volumes

Table numbers and table titles for all the County and County Borough volumes are identical. Page locations, however, vary from volume to volume. Hence, it is necessary to consult the Tables list and the Page Location Matrix which follow in order to locate specific data.

TABLES[1]

Number	Title
1.	Area, buildings for habitation and population, 1961.
2.	Population, 1821-1961.
3.	Population, 1951 and 1961, and intercensal changes.[2]
4.	Area, population, buildings for habitation, private households and valuation.[3]
5.	Area, population, buildings for habitation, private households and valuation.[4]
6.	Private dwellings: inhabited dwellings by building type, tenure and rooms.[5]

---

[1] Statistics for the county or county borough are included in every table except Table 11.

[2] All counties except Belfast and Londonderry: administrative areas.

[3] All counties except Belfast: administrative areas, district electoral divisions, towns. Belfast: wards.

[4] County electoral divisions. This table does not apply to Belfast.

[5] All counties except Belfast: administrative areas. Belfast: wards.

TABLE TITLES BY VOLUME

7. Private households: size, rooms occupied, and density of room occupation.[5]
8. Private households: availability of household arrangements by tenure and type of accommodation.[6]
9. Private households: households and persons therein by tenure of accommodation, rooms occupied and socio-economic group of head of household.
10. Inhabited buildings, etc. other than private dwellings: class, number, total population and (for institutions) number of inmates.[5]
11. Population, Parliamentary Electors, and Members of Parliament.[7]
12. Adjustment of enumerated (de facto) population to obtain resident (de jure) population.[6]
13. Ages by single years, sex and marital condition.
14. Ages by quinquennial groups, sex and marital condition.[5]
15. Birthplaces.
16. Birthplaces: persons born outside Northern Ireland by religion and place of birth.
17. Nationality: persons born outside the British Isles by nationality and usual address a year ago.
18. Religions.[5]
19. Religions: population under 22 years by individual years and 20 years and over by quinquennial groups.
20. Education: terminal education ages of persons 15 years and over by sex and age groups.
21. Education: terminal education ages of persons 15 years and over by sex, age group and occupation order.
22. Occupations: population aged 15 and over by occupation and industrial status.
23. Occupations: occupied population aged 15 and over by occupation, marital condition and age group.
24. Occupations: population aged 15 and over by socio-economic group, social class and age group.
25. Industries: working population aged 15 and over by industry (excluding persons out of work).

---

[6]All counties except Belfast: administrative areas.

[7]Parliamentary constituencies.

NORTHERN IRELAND

PAGE LOCATION MATRIX

Table Number	Belfast County Borough	County of Antrim	County of Armagh	County of Down	County of Fermanagh	County and County Borough of Londonderry	County of Tyrone
1	1	1	1	1	1	1	1
2	1	1	1	1	1	1	1
3	1	1	1	1	1	1	1
4	2	2	2	2	2	2	2
5	-	11	6	11	6	7	11
6	3	13	8	13	7	8	13
7	11	22	13	23	10	14	19
8	16	28	17	30	12	18	23
9	21	33	20	35	14	21	26
10	22	34	21	36	15	23	27
11	24	36	23	38	16	25	28
12	24	36	23	38	16	25	28
13	25	37	24	39	17	26	29
14	27	40	26	41	19	30	31
15	31	44	29	46	21	33	34
16	32	45	30	47	22	34	35
17	32	45	30	47	22	34	35
18	33	46	31	48	23	35	36
19	34	47	32	49	24	36	37
20	35	48	33	50	25	38	38
21	36	49	34	51	26	40	39
22	38	51	36	53	28	44	41
23	45	58	42	60	35	59	49
24	60	71	52	73	45	83	61
25	61	72	53	74	46	85	62

NORWAY

MAIN ENTRY AND VOLUME NOTES:[1]

Norway. Statistisk Sentralbyrå.
    Folketelling 1960. Population census 1960. Oslo, 1963-1964.
    8 v. 24cm. (66-33609)

Hefte I. Folkemengde og areal etter administrative inndelinger. Tettbygde strøk i herredene. Bebodde øyer. Volume I. Population and area by administrative divisions. Densely populated areas in rural municipalities. Inhabited islands. 1963. 236 p.

Hefte II. Folkemengden etter kjønn, alder og ekteskapelig status. Volume II. Population by sex, age and marital status. 1963. 167 p.

Hefte III. Folkemengden etter naering, stilling og sosial status. Volume III. Population by industry, occupation and status. 1964. 287 p.

Hefte IV. Utdanning. Volume IV. Education. 1964. 129 p.

Hefte V. Husholdninger og familiekjerner. Volume V. Households and family nuclei. 1964. 115 p.

Hefte VI. Boliger. Volume VI. Housing. 1964. 177 p. [Excluded.]

Hefte VII. Barnetallet i ekteskap. Volume VII. Fertility of marriages. 1964. 97 p.

Hefte VIII. Trossamfunn. Fødested. Statsborgerskap. Eiere av personbil. Leiligheter med telefon. Volume VIII. Religious denomination. Place of birth. Citizenship. Private car owners. Dwelling units with telephones. 1964. 85 p.

ENUMERATION DATE: November 30, 1960.

ENUMERATION BASIS: De jure.

---

[1] The census of Norway is published in two languages: Norwegian and English. The main entry and volume notes are given here in both languages; the table titles, in English only.

NORWAY

GLOSSARY

ADMINISTRATIVE, GEOGRAPHICAL, AND POLITICAL UNITS

Level	English Translation	Vernacular Term
National	Kingdom of Norway	Kongeriket Norge
Intermediate	Counties	Fylker or Fylkesoppgaver
	County districts	Fylkeskommuner
	Urban districts	Bykommuner
	Rural districts	Bygder
Local	Municipalities	Kommuner or Kommuneoppgaver
	Urban municipalities/Towns	Byer
	Rural municipalities	Herreder
	Suburban municipalities	Forstadskommuner

Population cluster ("locality") concepts:[1]

	Densely populated areas	Tettbygde strøk
	Urban areas	Bymessig bebygde strøk
	Suburbs [Remainder of densely populated areas]	Forstadskretser ...
	Sparsely populated areas	Spredtbygde strøk
	Non-urban areas	Ikke bymessig bebygde strøk

The Kingdom of Norway is divided into 20 <u>counties</u>. Of these, 18 are <u>county districts</u>, and 2, Oslo and Bergen, <u>urban districts</u>.

<u>Municipalities</u>, independent legal entities, are the units of local authority. There are two types: <u>urban municipalities</u>, also called <u>towns</u>, and <u>rural municipalities</u>, the latter directly responsible to the National Cabinet. Each urban municipality comprises an <u>urban district</u>; each rural municipality, a <u>rural district</u>. In addition to statistics for these divisions, the census further provides data for <u>suburban municipalities</u>, which are densely populated municipalities in which at least one-third of the economically active population is permanently employed in the nearest city.[2]

---

[1] For a discussion of this concept, see <u>European Population Censuses: The 1960 Series</u>, U.N. Doc.: ST/CES/3 (1964), p. 108.

[2] Samuel Humes and Eileen M. Martin, <u>The Structure of Local Governments Throughout the World</u> (The Hague: M. Nijhoff, 1961), p. 252.

GLOSSARY

Statistics are also given for areas delineated by density of population. Densely populated areas are groups of houses with at least 200 residents and a maximum of 50 meters between houses. Smaller clusters of houses which are integral parts of larger settled areas are included in the larger agglomerations, even though the distance from the clusters to the main areas is greater than 50 meters. Urban areas are densely populated areas with a residential population of at least 2,000 persons. Other densely populated areas are those containing fewer than 2,000 residents. Non-urban areas include the remainder of the densely populated areas (those parts not designated as urban areas or suburbs) plus the sparsely populated areas.

TABLE OF DEMOGRAPHIC CONCEPTS

The demographic concepts which appear in the following table correspond to the recommendations of the European Conference of Statisticians, unless otherwise noted. They are defined and discussed in the appendixes and in footnotes. Terms in parentheses are alternative forms which appear in the table titles.

Appendix	English Table Term	Vernacular Table Term
	ECONOMIC CHARACTERISTICS	
A	Economic activity	Økonomisk aktivitet
	Economically active	Yrkesbefolkningen
	Economically inactive	Ikke-inntekstakere
B	Employment status	Sosial status
C	Industrial classification	Naering
	Division(s) (Industry)	Naering
	Major group(s) (Industrial branch)	Naeringsgren
	Group(s) (Industrial group)	Naeringsgruppe
D	Occupational classification[1]	Stilling
	Major group(s) (Occupation 1-digit groups)	Stilling (1-siffernivå)
	Minor group(s) (Occupation 2-digit groups)	Stilling (2-siffernivå)
	Unit group(s) (Occupation 3-digit groups)	Stilling (3-siffernivå)
E	Socio-economic categories	Sosio-økonomisk gruppering

---

[1] In the table titles, "occupation (limited 3-digit groups)" [stilling (begrenset 3-siffernivå)] designates a tabulation of a selected list of the 3-digit level groups (i.e., the ISCO unit group level).

NORWAY

HOUSEHOLD AND FAMILY CHARACTERISTICS

F Households   Husholdninger
   Private households Private husholdninger
   Institutional households Felleshusholdninger

   Family nuclei   Familiekjerner

TABLE TITLES BY VOLUME

VOLUME 1. POPULATION AND AREA BY ADMINISTRATIVE DIVISIONS. DENSELY POPULATED AREAS IN RURAL MUNICIPALITIES.

TABLES

Number	Title	Page
1.	Main administrative divisions at 1 November 1960, by counties.[1]	8
2.	Population and area by municipalities.	10
3.	Resident population in areas affected by changes of municipality boundaries during the period 1 December 1950-1 November 1960.	27
4.	Persons temporarily present or temporarily absent from their residence, by municipalities.	29
5.	Population, members of the National Church and area in the ecclesiastical divisions.	38
6.	Resident population in the judicial divisions.	80
7.	Resident population in the primary court districts and the rural police districts.	81
8.	Population and area in the public health districts.	89
9.	Densely populated areas in rural municipalities.	100
10.	Towns and suburbs with resident population of 10,000 or more.	135
11.	Resident population, private households and area of inhabited islands.	149
12.	Resident population 1801-1960, by municipalities.	184

---

[1] The main administrative divisions of the counties fall into three categories: civil (towns, rural municipalities, etc.), ecclesiastical (deaneries, clerical districts, parishes, etc.), and judicial (court districts, police districts, public health districts, etc.).

TABLE TITLES BY VOLUME

VOLUME 2. POPULATION BY SEX, AGE, AND MARITAL STATUS.

TABLES

Number	Title	Page
1.	Population by sex and age. Municipalities.	8
2.	Persons 15 years and over, by sex and marital status. Municipalities.	74
3.	Population by sex, age, and marital status. Counties.	92
4.	Population by sex, age, and marital status. Detailed data for densely and sparsely populated areas.	106
5.	Population by sex and age in densely populated areas of rural municipalities.	124
6.	Population by sex and age in towns and suburbs with total resident population of 10,000 and over.	163

VOLUME 3. POPULATION BY INDUSTRY, OCCUPATION, AND STATUS.

TABLES

Number	Title	Page
1.	Resident population by sex, status in the household, and supporter's industrial branch and occupation (2-digit groups). The whole country.	16
2.	Resident population by industry. Municipalities.	22
3.	Persons 15 years and over by sex, economic activity, industry, employment status, and occupation (1-digit groups). Municipalities.	40
4.	Economically active population by sex, industrial branch, and employment status, and by sex and industrial group. Other persons 15 years and over by economic activity. Counties.	108
5.	Economically active population by sex, age, and marital status. Counties.	152
6.	Economically active population by sex, age, and industrial group. The whole country.	160
7.	Economically active population by sex, industrial group, and employment status. The whole country, rural districts, and towns.	172
8.	Economically active population in suburban municipalities by sex, industry, employment status, and work place.	191
9.	Economically active population by industry in towns	

NORWAY

	and suburbs with a resident population of 10,000 and over.	211
10.	Economically active population by sex, age, and occupation (1-digit groups). Counties.	215
11.	Economically active population by sex, age, and occupation (3-digit groups). The whole country.	220
12.	Economically active population by sex, occupation (2-digit groups), and employment status. The whole country, rural districts, and towns.	232
13.	Craftsmen by sex, occupation (3-digit groups), and employment status. The whole country, rural districts, and towns.	238
14.	Economically active population by sex, industrial branch and occupation (limited 3-digit groups). The whole country, rural districts, and towns.	242
15.	Persons 15 years and over with or without a secondary occupation by sex, main occupation, and employment status. The whole country, rural districts, and towns.	255

VOLUME 4. EDUCATION.[1]

TABLES

Number	Title	Page
1.	Persons 15 years and over by sex and highest general education with or without vocational education. Rural districts, towns by counties.	14
2.	Persons 15 years and over by sex, age, and general education. Rural districts, towns by counties.	16

---

[1]Questions dealt both with general education after public (grade) school and special (vocational) education. With respect to general education, persons were asked to indicate only the highest examination they had taken: (1) secondary school (upper level--gymnasium), (2) intermediate or secondary school (lower level), (3) continuation school (not gymnasium). Persons with vocational education were asked to list all degrees or certificates they had obtained which required the equivalent of at least one year of school. Where several types of vocational education were listed, only the "highest" and "second highest" were tabulated. The criteria for evaluating "highest" and "second highest" vocational education were primarily the length of time needed for training and the occupation of the informant.

298

3.	Persons 15 years and over by sex, occupation (2-digit groups), and general education. The whole country.	24
4.	Persons 15 years and over by sex, highest and second highest vocational education. The whole country.	27
5.	Persons 15 years and over by highest vocational education. Municipalities.	30
6.	Persons 15 years and over by sex, age, and highest vocational education. Municipality groups by district.	62
7.	Females 15 years and over by economic activity, age, and highest vocational education. Municipality groups by district.	84
8.	Females 15 years and over by economic activity, age, marital status, and highest vocational education. The whole country.	102
9.	Persons 15 years and over by sex, general education and highest vocational education. The whole country.	106
10.	Economically active population by sex, industrial branch and highest vocational education. Other persons 15 years and over by sex, economic activity, and highest vocational education. The whole country.	108
11.	Persons 15 years and over by sex, age, and educational level. Municipality groups by district.	120
12.	Pupils/students of secondary schools, vocational schools, and universities, etc. by sex and age. Number and percentages. Counties.	126

VOLUME 5. HOUSEHOLDS AND FAMILY NUCLEI.

TABLES

Number	Title	Page
1.	Households by type. Municipalities.	12
2.	Institutional households by type and present-in-area population by sex and status in the household. The whole country, rural districts, and towns.	29
3.	Institutional households by type, size, and present-in-area population by status in the household. The whole country, rural districts, and towns.	30
4.	Resident and present-in-area population in institutional households. Rural districts and towns by counties.	32

NORWAY

5.	Private households by size, and by sex, age, and marital status of head of household. The whole country, rural districts, and towns.	33
6.	Private household by size, and by sex and age of head of household. Rural districts and towns by districts.	36
7.	Private households by type and size. Rural districts and towns by counties.	41
8.	Private households by type, and by sex, age, and occupation (socio-economic groups) of head of household. The whole country.	57
9.	Private households by type, size, and by sex, age, and occupation (socio-economic groups) of head of household. The whole country.	64
10.	Private households by type, and by occupation (socio-economic groups) of head of household and number of dependents. The whole country.	76
11.	Private households by type, size, and by occupation (socio-economic groups) of head of household and number of dependents. The whole country.	80
12.	Private households by type, number of income earners and persons related to head of household. The whole country.	86
13.	Family nuclei by size, type of household, and type of family. The whole country, rural districts, and towns.	89
14.	Family nuclei by number of unmarried children, and by sex and age of head of the family nucleus. Rural districts and towns by district.	91
15.	Family nuclei by number of unmarried children in age groups. The whole country, rural districts, and towns.	96
16.	Persons in private households by sex, marital status, and status in the household. Rural districts and towns by counties.	100
17.	Family households, in which both spouses are economically active, by occupation of husband and wife. The whole country, rural districts, and towns.	111
18.	Family households, in which both spouses are economically active, by industry of husband and wife. The whole country, rural districts, and towns.	112
19.	Family households, in which both spouses are economically active, by education of husband and wife. The whole country, rural districts, and towns.	114

TABLE TITLES BY VOLUME

VOLUME 7. FERTILITY OF MARRIAGES.[1]

TABLES

Number	Title	Page
1.	Marriages by duration of marriage and by number of children. The whole country, urban, and non-urban areas.	14
2.	Marriages by age of wife at marriage, by duration of marriage, and by number of children. The whole country.	20
3.	Marriages and number of children, by age of wife at marriage, and by duration of marriage. The whole country, urban, and non-urban areas.	46
4.	Average number of children in marriages, by duration of marriage, and by age of wife at marriage. Rural districts by counties, Oslo, Bergen, Trondheim, and other towns.	58
5.	Marriages and number of children in marriages of duration 18 years and over, by occupation of husband, and by age of wife at marriage. The whole country, urban, and non-urban areas.	70
6.	Average number of children in marriages of duration 0-17 years, by occupation of husband, and by age of wife at marriage. The whole country.	74
7.	Marriages and number of children, by education of wife, by age of wife at marriage, and by duration of marriage. The whole country.	80
8.	Total marriages by age of wife and by number of children. The whole country, urban, and non-urban areas.	84
9.	Total marriages by age and economic activity of wife, and by number of children at home (by age of the youngest child). The whole country.	86
10.	Total marriages by duration of marriage, by number of children, and by economic activity of wife. The whole country.	88
11.	First births in marriages, by time between year of marriage and birth, by age of wife at marriage, and by year of marriage. The whole country.	91

---

[1] Data on marital fertility derive from a question on the number of children born alive in the existing marriage. The year of the existing marriage was also ascertained.

NORWAY

12. Births (other than first births) in marriages, by time between births, and by year of marriage. The whole country.   96

VOLUME 8. RELIGIOUS DENOMINATION.[1] PLACE OF BIRTH.[2] CITIZENSHIP. PRIVATE CAR OWNERS. DWELLING UNITS WITH TELEPHONES.

TABLES

Number	Title	Page
1.	Persons not belonging to the National Church, by religious community. Rural districts and towns by counties.	12
2.	Resident population by sex, age, and religious community. The whole country, rural districts, towns, urban, and non-urban areas.	16
3.	Persons 15 years and over not belonging to the National Church, by sex, age, religious community, and education. The whole country, rural districts, towns, urban, and non-urban areas.	18
4.	Persons not belonging to the National Church, by religious community and occupation (socio-economic groups). The whole country, rural districts, towns, urban, and non-urban areas.	30
5.	Resident population by place of birth. Rural districts and towns by counties.	34
6.	Resident population by place of birth. Municipalities.	40
7.	Persons born abroad by sex and country of birth. The whole country, rural districts, and towns.	55
8.	Persons born abroad by sex, age, and country of birth. The whole country.	56
9.	Aliens by sex, age, marital status, industry, occupation (1-digit groups), and country of citizenship. The whole country.	60
10.	Persons 15 years and over owning private car, by	

---

[1] Religious categories are as follows: the National Church, Roman Catholic, Methodist, Baptist, Adventist, Free Lutheran Evangelical, Pentecostalist, other religious communities, and outside all religious communities.

[2] The place of birth is the mother's residence at the time of the respondent's birth, not the actual place of birth.

TABLE TITLES BY VOLUME

	status in the household. Municipality groups by counties.	62
11.	Persons 15 years and over owning private car, by sex and age. The whole country.	65
12.	Persons 15 years and over owning private car, by occupation (2-digit groups). The whole country.	66
13.	Private households and occupants in dwelling units with telephone installed. Rural districts and towns by counties, and municipalities with population 10,000 and over.	68
14.	Private households and occupants in dwelling units with telephone installed, by sex and age of head of household. The whole country.	71
15.	Private households and occupants in dwelling units with telephone installed, by type and size of household. The whole country.	72
16.	Private households and occupants in dwelling units with telephone installed, by occupation (socio-economic groups) of head of household. Municipality groups by district.	73

PORTUGAL

MAIN ENTRY AND VOLUME NOTES:[1]

Portugal. Instituto Nacional de Estatística.
 X [i.e., Décimo] recenseamento geral da população, no continente e ilhas adjacentes (às 0 horas de 15 de Dezembro de 1960). [Various places.] 1963-1964.
 7 v. 29cm. (NUC63-46990)

Tomo I. Prédios e fogos; população. Volume 1. Dados retrospectivos (distritos, concelhos e freguesias). [1964] 260 p. Volume 2. Dados retrospectivos (lugares). [1964] 1164 p.

Tomo II. Famílias, convivências e população residente e presente, por freguesias, concelhos, distritos e centros urbanos. [1963] 662 p.

Tomo III. Volume 1. Idade. [1963] 180 p. Volume 2. Instrução. [1964?] 195 p.

Tomo IV. Estrangeiros; órfãos; cegos e surdos-mudos. [1963] 173 p.

Tomo V. Condições perante o trabalho e meio de vida. Volume 1. Total geral; totais dos centros urbanos e das zonas rurais. [1964] 323 p. Volume 2. Distritos. [1964] 424 p. Volume 3. Concelhos e centros urbanos. [1964] 176 p.

Tomo VI. Condições de habitação dos agregados domésticos. [1964] 500 p.

[Tomo VII] Anexo. Inventário de prédios e fogos (em julho de 1960). [1962] 284 p. [Excluded.]

ENGLISH TRANSLATION OF CENSUS AND VOLUME TITLES:

Tenth general population census of the continent and adjacent islands as of 0 hours, December 15, 1960.

I. Buildings and households; population. Summary data.
 1. Districts, municipalities, and parishes.
 2. Hamlets.

II. Families, institutional households, and resident and

---

[1] The census of Portugal is published in two languages: Portuguese and French. The French volume titles, etc., are not included here.

enumerated population by parish, municipality, district, and urban center.
- III. [No volume title given; 2 subtitles, only, as follows:]
  1. Age.
  2. Education.
- IV. Foreigners, orphans, blind, and deaf-mutes.
- V. Economic activity.
  1. General total; urban centers (total) and rural zones (total).
  2. Districts.
  3. Municipalities and urban centers.
- VI. Living conditions of domestic households.
- [VII] Annex: Inventory of buildings and households (in July 1960).

ENUMERATION DATE: December 15, 1960.

ENUMERATION BASIS: De facto; de jure.

GLOSSARY

ADMINISTRATIVE, GEOGRAPHICAL, AND POLITICAL UNITS

Level	English Translation	Vernacular Term
National	Republic of Portugal	República portuguesa
Intermediate	Province	Província
	District	Distrito
Local	Municipality	Concelho
	Parish	Freguesia
Other concepts:		
	Urban center	Centro urbano
	Rural zone	Zona rural
	Hamlet	Lugar

For administrative purposes, Portugal is divided into 11 provinces, which have representative political organs. The provinces are made up of municipalities, and these in turn of parishes. Districts, likewise administrative arms of the central government, have no "legal personality."[1]

---

[1] Samuel Humes and Eileen M. Martin, The Structure of Local Governments Throughout the World (The Hague: Martinus Nijhoff, 1961), pp. 340-341.

Urban centers are the capitals or principal population agglomerations of territorial divisions. In a municipality, for example, the urban zone embraces 10,000 or more inhabitants.

Rural zones are those parts of a territorial division not included in the urban centers.

Hamlets are the smallest population aggregates in rural areas.

TABLE OF DEMOGRAPHIC CONCEPTS

The demographic concepts which appear in the following table correspond to the recommendations of the European Conference of Statisticians, unless otherwise noted. They are defined and discussed in the appendixes and in footnotes. Terms in brackets are not used in the table titles, but they are included here in order to give logical completeness to the various classifications in the table.

Appendix	English Table Term	Vernacular Table Term
CULTURAL, EDUCATIONAL, AND PERSONAL CHARACTERISTICS		
...	Religion[1]	Religião
ECONOMIC CHARACTERISTICS		
A	Economic activity	Condição perante o trabalho
	Active[2]	Activo
	With a profession[3]	Com profissão
	With an occupation[4]	Com ocupação
	Unemployed[5]	Desempregado
	In military service[6]	A prestar serviço militar
	Inactive[7]	Inactivo
B	Employment status	Situação na profissão
C	Industrial classification	Actividades económicas
	Division(s)	Divisões
	Major group(s)	Classes
	Group(s)	Grupos
D	Occupational classification[8]	Profissões
	Major group(s)	Grandes grupos
	[Minor group]	...
	Unit group(s)	Grupos de base
	Livelihood[9]	Meio de vida
E	Socio-economic category	Condição sócio-económica

---

[1][All footnotes are at the end of the Table.]

GLOSSARY

HOUSEHOLD AND FAMILY CHARACTERISTICS

F   Domestic household   Agregado domestico
        One-family   Unifamiliar
        Multi-family   Multifamiliar
        Family[10]   Família

---

[1] Categories of religion are as follows: Catholic, non-Catholic Christian, non-Christian, or without religion.

[2] The term "economically active" does not readily conform to the definition recommended by the European Conference of Statisticians. The category comprises persons 10 years of age or older as described in footnotes 3 through 6 below.

[3] "With a profession" are those persons who practice an occupation or work for remuneration.

[4] "With an occupation" are those persons who have a job without remuneration, or who live off income from investments or similar sources.

[5] "Unemployed" refers to those persons in search of a new or a first job.

[6] "In military service" refers to those persons conscripted for military service.

[7] In defining economically inactive, no explicit division is made between the independent and the dependent. Moreover, those who are living off income from investments or similar sources are included in the economically active population. Economically inactive includes the following categories: (1) retired persons; (2) invalids; (3) students without profession or occupation who are pursuing studies at any level [they may be dependent or independent]; (4) persons otherwise inactive who cannot be included in an active or any other inactive category [e.g., prisoners or permanent inmates in mental institutions].

[8] In the table titles, "occupation" designates tabulations of only major and unit groups; there are no tabulations of minor groups. "Occupation" in this sense should not be confused with the usage denoting the economic activity of unpaid workers and those living off income from investments or similar sources (see footnote 4 above).

[9] The respondent is classified by livelihood (principal source of support) in one of two categories: (1) "independent" if his support comes primarily from remuneration for work, property revenues, or pensions; (2) "dependent" if his support comes from

PORTUGAL

internment in a public or private welfare institution, assistance without internment, charities, heads of families, or other persons.

[10] The definition of family is based on the concept of related members of a household, not on the concept of a family nucleus. (See Tome II below.)

TABLE TITLES BY VOLUME

TOME I, VOLUME 1. BUILDINGS AND HOUSEHOLDS; POPULATION. SUMMARY DATA (DISTRICTS, MUNICIPALITIES, AND PARISHES).

TABLES

Number	Title	Page
1.	Buildings and households according to the 1940, 1950, and 1960 censuses, by district and municipality.	3
2.	Resident population at successive censuses, 1864-1960, by parish.	11
3.	Density of the resident population at successive censuses, 1864-1960, by municipality.	105

TOME I, VOLUME 2. BUILDINGS AND HOUSEHOLDS; POPULATION. SUMMARY DATA (HAMLETS).

TABLES

Number	Title	Page
1.	Buildings according to the 1960 census; households and population at the 1911, 1940, and 1960 censuses, by hamlet.	3
2.	Hamlets by number of inhabitants (resident population), by municipality.	614
3.	Percentage distribution of the population by size of hamlet, by municipality. Censuses of 1911, 1940, and 1960.	632

TOME II. FAMILIES,[1] INSTITUTIONAL HOUSEHOLDS, AND RESIDENT AND ENUMERATED POPULATION BY PARISH, MUNICIPALITY, DISTRICT, AND URBAN CENTER.

TABLES

Number	Title	Page
1.	Families, institutional households, and resident	

---

[1] A family is a group of persons united by bonds of kinship,

	population by sex, marital status, and religion,[2] showing the number of temporarily absent individuals, by district, municipality, and parish.	2
2.	Families by number of persons and type of family, by district and municipality.	218
3.	Institutional households by type of household and number of persons present by household type and the household's social function (educational, health, religious, etc.), by district and municipality.	292
4.	Resident population having changed residence in 1960 by former residence, by district and municipality.	308
5.	Naturalized Portuguese by sex and origin and foreigners by sex, by district and municipality.	316
6.	Resident population by age (five-year groups) and sex, by district and municipality.	332
7.	Resident population by religion[2] and sex, by district and municipality.	364
8.	Enumerated population by sex and marital status, showing the number of non-residents present, by district, municipality, and parish.	372
9.	Non-resident enumerated population by usual residence, by district and municipality.	588
10.	Families, institutional households, and resident population by sex, marital status, and religion,[2] showing the number of temporarily absent individuals, in urban centers (total) and individual urban centers.	596
11.	Enumerated population by sex and marital status, showing the number of non-residents present, in urban centers (total) and individual urban centers.	600

---

legitimate or illegitimate, habitually using the same dwelling, or a solitary person occupying a dwelling. Families are divided into the following types: (1) couples without children; (2) couples with children, without other relatives; (3) couples with children and other relatives; (4) persons with children, without other relatives; (5) persons with children and other relatives; (6) other persons; (7) solitary persons.

[2] Catholic, other religion, or without religion.

PORTUGAL

TOME III, VOLUME 1. AGE.[1]

TABLES

Number	Title	Page
1.	Resident population by sex, marital status, and single years of age. General total (continent and islands).	2
2.	Resident population by sex, marital status, and age (five-year groups), in urban centers, rural zones (total), and districts.	8
3.	Ever-married women by marital status, age at last marriage, and age at the date of the census, with figures showing the number of children born. General total (continent and islands), urban centers (total), and rural zones (total).	118
4.	Ever-married women who have borne children within the past five years, by marital status, age at last marriage, and age at the date of the census, with figures showing the number of children which they have borne. General total (continent and islands), urban centers (total), and rural zones (total).	124
5.	Ever-married women who have borne children within the past year, by marital status, age at last marriage, and age at the date of the census, with figures showing the number of children which they have borne. General total (continent and islands), urban centers (total), and rural zones (total).	130
6.	Resident population by religion, sex, and age (five-year groups), in urban centers (total), rural zones (total), and districts.	136

TOME III, VOLUME 2. EDUCATION.[2]

TABLES

Number	Title	Page
1.	Resident population by type of education; resident	

---

[1] Age is defined as the number of complete years lived from the date of birth to the census date.

[2] Education comprises the following four categories: pre-school, primary, secondary, and higher education.

population, not attending school, by level of education, specialty, age,[3] and sex. General total for the continent and islands. ... 1

2. Resident population by type of education; resident population, not attending school, by level of education, specialty, and age,[4] in urban centers, all rural zones, and districts. ... 6

3. Population attending school by level of instruction, specialty, age,[3] and sex. General total for the continent and islands. ... 94

4. Population attending school by level of education, specialty, and age,[5] in urban centers, all rural zones, and districts. ... 98

5. Resident population 7 years of age and older by type and level of education and sex, in districts and municipalities. ... 142

TOME IV. FOREIGNERS, ORPHANS, BLIND, AND DEAF-MUTES.

TABLES

Number	Title	Page

Part One: Foreigners

1. Resident foreigners by nationality, in urban centers (total), rural zones (total), by districts and municipalities. ... 22

2. Resident foreigners by sex, marital status, and duration of stay, showing the number of non-residents by nationality, in urban centers (total), rural zones (total), and the cities of Lisbon and Porto. ... 36

3. Resident foreigners by age, level of education, religion, and nationality, in urban centers (total), rural zones (total), and the cities of Lisbon and Porto. ... 40

4. Resident foreigners, 10 years of age or older, by nationality and economic activity, in urban centers

---

[3] Single years of age through 29 years, thereafter by five-year groups.

[4] Five-year groups.

[5] Five-year groups through 39 years, then 40 years and over as one group.

	(total), rural zones (total), and the cities of Lisbon and Porto.	44
5.	Resident foreigners by nationality, showing their means of livelihood, in urban centers (total), rural zones (total), and the cities of Lisbon and Porto.	50
6.	Resident foreigners by nationality, showing their socio-economic condition, in urban centers (total), rural zones (total), and the cities of Lisbon and Porto.	54
7.	Economically active resident foreigners (having a profession), 10 years of age or older, by nationality, industrial division, major group, and group, in urban centers (total), rural zones (total), and the cities of Lisbon and Porto.	62
8.	Economically active resident foreigners (having a profession), 10 years of age or older, by nationality and occupational major and minor groups, in urban centers (total), rural zones (total), and the cities of Lisbon and Porto.	70

Part Two: Orphans

9.	Orphans aged less than 18 years by sex, age, and type of orphanhood, in urban centers (total), rural zones (total), districts, and municipalities.	119
10.	Orphans aged less than 18 years by means of livelihood and socio-economic category, in urban centers (total), rural zones (total), districts, and the cities of Lisbon and Porto.	126

Part Three: Physical Defectives

11.	The blind by sex, age, and level of education, in urban centers (total), rural zones (total), districts, and municipalities.	131
12.	The blind by means of livelihood and socio-economic category, in urban centers (total), rural zones (total), districts, and the cities of Lisbon and Porto.	138
13.	Deaf-mutes by sex, age, and level of education, in urban centers (total), rural zones (total), districts, and municipalities.	141
14.	Deaf-mutes by means of livelihood and socio-economic category, in urban centers (total), rural zones (total), districts, and the cities of Lisbon and Porto.	148

TABLE TITLES BY VOLUME

TOME V, VOLUME 1. ECONOMIC ACTIVITY (GENERAL TOTAL; URBAN CENTERS [TOTAL] AND RURAL ZONES [TOTAL]).

TABLES

Number	Title	Page
1.	Resident population, 10 years of age and older, by economic activity, sex, and age, in urban centers (total) and rural zones (total).	2
2.	Economically active resident population (having a profession) by age and unemployed population seeking new employment; disabled persons by cause of disability; and those in military service; by occupation and sex, in urban centers (total) and rural zones (total).	4
3.	Economically active resident population (having a profession) by level of education, sex, and occupation, in urban centers (total) and rural zones (total).	56
4.	Economically active resident population (having a profession) by age, level of education, employment status, and sex, showing the number of family heads and their dependents, in urban centers (total) and rural zones (total).	92
5.	Economically active resident population (having a profession) by employment status, occupation, and sex, showing the number of family heads and their dependents, in urban centers (total) and rural zones (total).	98
6.	Economically active resident population (having a profession) by employment status, sex, division, major group, and group in industry, in urban centers (total) and rural zones (total).	146
7.	Economically active population (having a profession) by age, sex, and division, major group, and group of industry, showing the number of family heads and their dependents, in urban centers (total) and rural zones (total).	190
8.	Economically active population (having a profession) by division, major group, and group of industry, occupation, and sex, in urban centers (total) and rural zones (total).	226
9.	Economically active resident population (having an occupation) by occupation, sex, and age, in urban centers (total) and rural zones (total).	282
10.	Unemployed and disabled population and population in	

	military service by sex and employment status, in urban centers (total) and rural zones (total).	283
11.	Unemployed and disabled population and population in military service by sex and major group of industry, in urban centers (total) and rural zones (total).	285
12.	Resident population by age, socio-economic category, and sex, in urban centers (total) and rural zones (total).	290
13.	Resident population by age, means of livelihood, and sex, showing the number of family heads and their dependents, in urban centers (total) and rural zones (total).	296
14.	Resident population by means of livelihood, sex, and economic activity, in urban centers (total) and rural zones (total).	300

TOME V, VOLUME 2. ECONOMIC ACTIVITY (DISTRICTS).

TABLES

Number	Title	Page
1.	Resident population, 10 years of age and older, by economic activity, sex, and age, in districts and the cities of Lisbon and Porto.	2
2.	Economically active resident population (having a profession) by occupation and sex, in districts and the cities of Lisbon and Porto.	18
3.	Economically active resident population (having a profession) by age, level of education, employment status, and sex, showing the number of family heads and their dependents, in districts and the cities of Lisbon and Porto.	38
4.	Economically active resident population (having a profession) by age, sex, and division, major group, and group of industry, showing the number of family heads and their dependents, in districts and the cities of Lisbon and Porto.	82
5.	Economically active resident population (having an occupation) by classification of occupation and sex, in districts and the cities of Lisbon and Porto.	343
6.	Resident population by socio-economic category and sex, in districts and the cities of Lisbon and Porto.	344
7.	Resident population by age, means of livelihood, and sex, showing the number of family heads and their dependents, in districts and the cities of Lisbon and Porto.	346

8. Resident population by means of livelihood, sex, and economic activity, in districts and the cities of Lisbon and Porto. ... 386

TOME V, VOLUME 3. ECONOMIC ACTIVITY (MUNICIPALITIES AND URBAN CENTERS).

TABLES

Number	Title	Page
1.	Resident population, 10 years of age and older, by economic activity and sex, by municipality and urban center.	2
2.	Economically active resident population (having a profession) by occupational major group and sex, by municipality and urban center.	20
3.	Economically active resident population (having a profession) by employment status and sex, by municipality and urban center.	38
4.	Economically active resident population (having a profession) by industrial division, major group, group, and sex, by municipality and urban center.	56
5.	Economically active resident population (having an occupation) by occupation and sex, by municipality and urban center.	96
6.	Resident population by socio-economic category and sex, by municipality and urban center.	106
7.	Resident population by means of livelihood and sex, by municipality and urban center.	144

TOME VI. LIVING CONDITIONS OF DOMESTIC HOUSEHOLDS.[1]

TABLES

Number	Title	Page
1.	Domestic households by kind of building, type of dwelling, and type of household, in urban centers	

---

[1] A domestic household is a group of persons living on the same premises for reasons of family life, including servants and other persons living in, who need not be kin. The two types of domestic households are: (1) single family, i.e., the persons constituting the household form only one family, and (2) multi-family, i.e., the persons constituting the household form two or more families.

	(total), rural zones (total), districts, municipalities, and individual urban centers.	3
2.	Domestic households by number of persons and type of household, in urban centers (total), rural zones (total), districts, municipalities, and individual urban centers.	40
3.	Domestic households with dwelling in a building by form of occupancy and type of household, in urban centers (total), rural zones (total), districts, municipalities, and individual urban centers.	112
4.	Domestic households with dwelling in a building by number of rooms and type of household, in urban centers (total), rural zones (total), districts, municipalities, and individual urban centers.	148
5.	Domestic households with dwelling in a building by number of persons, number of rooms, and type of household, in urban centers (total), rural zones (total), and districts.	222
6.	Multi-family domestic households with dwelling in a building by characteristics of the dwelling, in urban centers (total), rural zones (total), districts, municipalities, and individual urban centers.	340
7.	Single family domestic households with dwelling in a building by characteristics of the dwelling, in urban centers (total), rural zones (total), districts, municipalities, and individual urban centers.	350
8.	Domestic households by socio-economic category of head and type of household, in urban centers (total), rural zones (total), and districts.	360
9.	Single family domestic households with dwelling in a building by number of rooms of the dwelling and socio-economic category of head, in urban centers (total), rural zones (total), and districts.	368
10.	Single family domestic households with dwelling in a building by characteristics of the dwelling and socio-economic category of head, in urban centers (total), rural zones (total), and districts.	446

SCOTLAND

MAIN ENTRY AND VOLUME NOTES:[1]

Scotland. General Register Office.
   Census 1961: Scotland. Edinburgh, 1963-1966.
   10 v. 34cm.

Volume 1. County report[s].
  Part 1. City of Edinburgh. 1963. 80 p.
  Part 2. City of Glasgow. 1963. 109 p.
  Part 3. City of Aberdeen. 1963. 62 p.
  Part 4. City of Dundee. 1963. 58 p.
  Part 5. County of Aberdeen. 1963. 103 p.
  Part 6. County of Angus. 1963. 110 p.
  Part 7. County of Argyll. 1964. 84 p.
  Part 8. County of Ayr. 1964. 148 p.
  Part 9. County of Banff. 1963. 80 p.
  Part 10. County of Berwick. 1964. 68 p.
  Part 11. County of Bute. 1964. 64 p.
  Part 12. County of Caithness. 1963. 66 p.
  Part 13. County of Clackmannan. 1964. 72 p.
  Part 14. County of Dumfries. 1964. 102 p.
  Part 15. County of Dunbarton. 1964. 124 p.
  Part 16. County of East Lothian. 1963. 74 p.
  Part 17. County of Fife. 1964. 150 p.
  Part 18. County of Inverness. 1963. 98 p.
  Part 19. County of Kincardine. 1963. 68 p.
  Part 20. County of Kirkcudbright. 1964. 72 p.
  Part 21. County of Lanark. 1964. 141 p.
  Part 22. County of Midlothian. 1963. 84 p.
  Part 23. Counties of Moray and Nairn. 1963. 122 p.
  Part 24. County of Orkney. 1963. 76 p.
  Part 25. County of Peebles. 1964. 66 p.
  Part 26. Counties of Perth and Kinross. 1964. 148 p.
  Part 27. County of Renfrew. 1964. 130 p.
  Part 28. County of Ross and Cromarty. 1963. 82 p.
  Part 29. County of Roxburgh. 1964. 76 p.

---

[1]Data concerning scientific and technological qualifications (based on a 10 percent sample) are also available. They are included as part of the Great Britain census: Report on Scientific and Technological Qualifications, q.v.

SCOTLAND

Part 30. County of Selkirk. 1964. 72 p.
Part 31. County of Stirling. 1964. 114 p.
Part 32. County of Sutherland. 1963. 64 p.
Part 33. County of West Lothian. 1963. 88 p.
Part 34. County of Wigtown. 1964. 66 p.
Part 35. County of Zetland. 1963. 74 p.

Volume 2. Usual residence. 1965. xvi, 17 p.

Volume 3. Age, marital condition, and general tables. 1965. lxx, 109 p.

Volume 4. Housing and households.
  Part I. Housing and households. 1966. lxv, 193 p. [Excluded.]
  Part II. Household composition tables. 1966. xxxvi, 234 p.

Volume 5. Birthplace and nationality. 1966. xxv, 39 p.

Volume 6. Occupation, industry, and workplace.
  Part I. Occupation tables. 1966. xxxi, 199 p.
  Part II. Industry tables. 1966. xxxvi, 237 p.
  Part III. Workplace tables. 1966. xxi, 61 p.

Volume 7. Gaelic. 1966. xxiii, 28 p.

Volume 8. Internal migration. 1966. xliii, 127 p.

Volume 9. Terminal education age. 1966. xix, 23 p.

Volume 10. Fertility. 1966. xxiii, 289 p.

ENUMERATION DATE: April 23, 1961.

ENUMERATION BASIS: De facto.

GLOSSARY

ADMINISTRATIVE, GEOGRAPHICAL, AND POLITICAL UNITS

Level	Vernacular Term
National	Scotland
Intermediate	...

Local	Local authority areas:
	  County
	    Burgh
	      Large burgh
	      Small burgh
	    Landward area
	      District of county
	  County of city/city

GLOSSARY

Other concepts: Burghal/landward aggregates
                                Conurbation
                                    Conurbation center
                                New town
                                Regional division

Scotland is divided for administrative purposes into <u>local authority areas</u> comprising <u>counties</u> (and their subdivisions) and <u>counties of cities</u>.

Counties comprise <u>burghs</u> and <u>landward areas</u>. Burghs are municipal units under the jurisdiction of the county council. <u>Large burghs</u> have populations of 20,000 or more, and <u>small burghs</u> have populations of fewer than 20,000.[1]

<u>Landward areas</u> are the non-burghal parts of counties. In all counties except Kinross and Nairn, these are divided into <u>districts of counties</u>, each with its own district council.

<u>Counties of cities</u> (Edinburgh, Glasgow, Aberdeen, and Dundee) are analogous to the county boroughs in England and Wales, whose town councils are all-purpose authorities. Counties of cities is usually abbreviated to <u>cities</u> in the census.

<u>Burghal/landward aggregates</u> comprise (1) the conurbation, if applicable, and (2) areas outside the conurbation: (a) burghal areas with 100,000 or more population, (b) burghal areas with 10,000 but fewer than 100,000 population, (c) burghal areas with fewer than 10,000 population, and (d) landward areas.

A <u>conurbation</u> is a metropolitan center surrounded by an aggregation of urban communities. Central Clydeside Conurbation is the only conurbation in Scotland, and the <u>conurbation center</u> is its central area. This area, the Exchange Ward of the city of Glasgow, is characterized by a relatively small residential population and a large concentration of commuting workers.

<u>New towns</u>, of which there are 3 in Scotland, were created by the New Towns Act in 1946 to encourage the decentralization of population and industry concentrated in the large cities. The Act specified relatively underdeveloped locations for the self-contained and locally governed new towns, which form part of the landward areas of counties.

<u>Regional divisions</u> and subdivisions are geographical groupings of counties and cities for purposes of census tabulation.

---

[1] Samuel Humes and Eileen M. Martin, <u>The Structure of Local Governments Throughout the World</u> (The Hague: Martinus Nijhoff, 1961), pp. 212-214.

SCOTLAND

TABLE OF DEMOGRAPHIC CONCEPTS

The demographic concepts which appear in the following table correspond to the recommendations of the European Conference of Statisticians, unless otherwise noted. They are defined and discussed in the appendixes and in footnotes. Terms in brackets are not used in the table titles, but they are included here in order to give logical completeness to the various classifications in the table.

Appendix	English Table Term	United Kingdom Terminology
CULTURAL, EDUCATIONAL, AND PERSONAL CHARACTERISTICS		
...	...	Birthplace and citizenship or nationality[1]
ECONOMIC CHARACTERISTICS		
A	Economic activity	Same
	Economically active	Same
	Economically inactive	Same
	Earners[2]	Same
B	Employment status	Same
C	Industrial classification[3]	Same
	Division(s)	Orders
	Major group(s)	Orders
	Group(s)	Minimum list headings
D	Occupational classification[4]	Same
	Major group(s)	Orders
	Minor group(s)	Unit groups
	[Unit group][5]	...
E	Socio-economic category[6]	Socio-economic group
	...	Status group[7]
HOUSEHOLD AND FAMILY CHARACTERISTICS		
F	Household	Same
	Private household	Same
	Types of household[8]	Same
	Head of household[9]	Same
	Chief economic supporter of household[10]	Same
	Institutional household	Same
	Family	Same
	Type of family[11]	Same
	Head of family[12]	Same

[1] [All footnotes for the Table are on the following pages.]

GLOSSARY

[1] The classification by birthplace separates Scotsmen who were born in Edinburgh, Glasgow, Dundee, or Aberdeen City and those born elsewhere in Scotland. The birthplace of the latter is given by county. For those born outside Scotland, the classification is by country of birth, e.g., England, Wales, Northern Ireland, Poland.

For those born outside Great Britain and Northern Ireland the census schedule posed three questions concerning citizenship and nationality: (1) if a citizen of the Commonwealth, state citizenship, e.g., United Kingdom and Colonies, Indian, Canadian; (2) if a citizen of the United Kingdom and Colonies, state whether citizen by birth, descent, naturalization, etc.; (3) if other than (1) or (2), state nationality, e.g., Italian, Polish.

[2] In a household an earner is any economically active person, including domestic servants.

[3] In the table titles, "industry" designates tabulations of any and all levels of industrial classification. The U.K. system of industrial classification corresponds to the ISIC 1958 as follows: "Orders" are "Divisions" and "Major groups"; "Minimum list headings" are "Groups."

[4] In the table titles, "occupation" designates tabulations of either the ISCO 1958 major and minor groups together, or of the major groups alone. In the U.K. system of occupational classification, "Orders" are "Major groups," "Unit groups" are "Minor groups."

[5] U.K. use of the term "Unit group" is not to be confused with the ISCO 1958 definition.

[6] The classification by socio-economic group is similar, but not identical, to that recommended by the European Conference of Statisticians, described in Appendix E. Full definitions of the socio-economic groups in terms of occupation and employment status are given in Great Britain, <u>Classification of Occupations, 1960</u>.

[7] The economically active and inactive populations are divided into status groups, which are defined by combining classifications by employment status and economic position.

<u>Economically active</u>
1. Out of employment
    a. Sick
    b. Others
2. In employment
    a. Self-employed
        i. Without employees
        ii. With employees
            1. Large establishments
            2. Small establishments

         b. Managers
              i. Large establishments
              ii. Small establishments
         c. Foremen and supervisors
         d. Apprentices, articled clerks and formal trainees
         e. Professional employees
         f. Other employees
         g. Family workers
         h. Part-time workers
Economically inactive
    1. Institution inmates
    2. Retired persons
    3. Students
    4. Other persons economically inactive--all persons over the age of 15 without paid occupation or former occupation, including housewives.

[8] The principal types of households are as follows: "0" - No family; "1" - One family; "2" - Two families; "3" - Three or more families. A detailed description of these types and their numerous subdivisions can be found in Appendix 1 to Volume 4, Part II of the census (pp. xxii-xxvi).

[9] The head of the household is the person so described on the census schedule by the householder making the return.

[10] The chief economic supporter is selected from among those members of the household who are not boarders, or employees of the head of the household, or unrelated to the head of the household, or under the age of 15. The chief is selected by applying the following set of rules:

   a. Employment status is considered first. Those employed full time (or whose hours worked per week are not stated), or out of employment, are selected before those employed part time, who are selected before those retired, who are selected before any others.

   b. Among those selected above, position in the family is considered next, family heads being selected before any other member of the family, or before persons not in the family.

   c. Among those selected by rules (a) and (b), sex is considered next, males being selected before females.

   d. Among those selected by rules (a), (b), and (c), age is considered next, older persons being selected before younger.

If two or more people are finally selected according to these rules, that person whose name appears first on the census schedule is selected as chief.

TABLE TITLES BY VOLUME

[11]A family is a married couple, alone, or with their never-married child or children (of any age). A family may also be a lone parent with his/her never-married child or children. A lone parent is a married parent whose spouse does not reside in the same household, or any single, widowed, or divorced parent. The term child includes stepchild or adopted child (but not foster child) and also grandchild (without parents) or great-grandchild (without parents or grandparents).

Where numbers of children are tabulated in the body of a table, or where reference is made to children in row or column headings other than descriptive titles of household types, these children are dependent children, defined as children under the age of 15, or as persons of any age in full-time education.

For the purpose of defining a family or a household type, however, a child is a never-married person of any age (including stepchildren or adopted children) living in the same household as at least one of his/her parents, or at least one of his/her grandparents when there are no parents, or at least one of his/her great-grandparents when there are no parents or grandparents.

[12]The head of a family is the husband in the case of a married couple; otherwise, he/she is the lone parent.

TABLE TITLES BY VOLUME

VOLUME 1. COUNTY REPORT[S].
[35 Parts; see listing on pp. 317-318 above.]

Structure of the County Reports

Table numbers and table titles for all the County Reports are identical, but page locations vary from Part to Part. Hence it is necessary to consult the Tables list as well as the Page Location Matrix which follow in order to locate desired data.

TABLES

Number	Title
1.	Population 1801-1961 and intercensal variations.
2.	Population 1931-1961 and intercensal variations.[1]
3A.	Acreage, population, private households and dwellings.[2]

[1]Parts 5-35: County, large burghs, small burghs, districts of county.

[2]Parts 1-4: County of city, city wards. (Part 2: add Conurbation center.) Parts 5-35: County, large burghs, small burghs,

SCOTLAND

3B. Population by sex.[3]
3C. Population, 1951 and 1961.[4]
3D. Population, electorate.[5]
4A. Population, historical table.[6]
4B. Population, historical table, 1801-1961.[3]
5. Intercensal changes of boundary, 1951-1961.[7]
6. Age and marital condition.[8]
7. Age--single years under 21.[8]
8. Birthplaces and nationalities of the whole population.[9]
9. Residents born outside Scotland by nationality and citizenship, and visitors from outside Scotland.[8]
10. Birthplaces, nationalities and citizenships of residents of Scotland born outside the British Isles.
11. Dwellings by building type, rooms and household spaces.[8]
12. Buildings and dwellings.[10]
13. Private households by size, rooms occupied and sharing of dwellings.[8]

---

burgh wards, districts of county, county council electoral divisions, new towns.

[3] Parts 1-35: Civil parishes.

[4] Parts 1-35: Inhabited islands.

[5] Parts 1-35: Parliamentary constituencies.

[6] Parts 5-35: Large burghs, small burghs.

[7] Parts 1-35: Areal changes of any county subdivisions during the intercensal period are made explicit in Table 5 of each County Report.

[8] Parts 1-4: County of city, city wards. (Part 2: add Conurbation center.) Parts 5-35: County, large burghs, small burghs, districts of county, new towns.

[9] Parts 5-35: County, large burghs, small burghs with populations of 10,000 or more, county remainder, new towns.

[10] Part 2: County of city, conurbation center. Parts 5-31, 33, 34: County, large burghs, small burghs with populations of 10,000 or more, county [burghal/landward] aggregates, new towns. Parts 32, 35: County, county [burghal/landward] aggregates.

TABLE TITLES BY VOLUME

14. Private households by size, rooms occupied and type of building.[10]
15. Dwellings by tenure and rooms.[10]
16. Private households by size, rooms occupied, sharing of dwelling and tenure.[10]
17. Private households, persons and rooms by tenure.[11]
18. Private households by density of occupation (persons per room).[8]
19. Population in all private households by density of occupation (persons per room).[8]
20. Private households by type of building and tenure (proportions per 1,000).[12]
21. Households sharing a dwelling as a percentage of all households in each tenure category.[10]
22. Dwellings by availability of certain household arrangements.[8]
23. Private households by availability of certain household arrangements.[8]
24. Population enumerated in private and non-private households.[8]
25. Institutions.[8]
26. Hotels and boarding houses. Number, rooms and population.[13]
27. One- and two-person households containing persons of pensionable age by sex/condition combination.[8]

---

[11] Parts 1-4: City wards. Parts 5-35: Small burghs with populations of less than 10,000, districts of county.

[12] Parts 1-4: County of city, city wards. (Part 2: add Conurbation center.) Parts 5-31, 33, 34: County, large burghs, small burghs with populations of 10,000 or more, county [burghal/landward] aggregates, new towns. Parts 32, 35: County, county [burghal/landward] aggregates.

[13] Parts 1-4: County of city, city wards. (Part 2: add Conurbation center.) Parts 5-35: County, large burghs, small burghs, and districts of county with 500 or more hotel, etc. rooms, county remainder.

SCOTLAND

PAGE LOCATION MATRIX

Table Number	City of Edinburgh	City of Glasgow	City of Aberdeen	City of Dundee	County of Aberdeen	County of Angus	County of Argyll	County of Ayr	County of Banff	County of Berwick	County of Bute	County of Caithness
1	19	19	19	17	19	19	19	19	19	19	19	19
2	20	20	20	18	20	20	20	20	20	20	20	20
3A	20	20	20	18	21	21	21	21	21	21	20	20
3B	21	21	21	19	23	22	22	23	22	22	21	21
3C	21	23	21	19	25	24	24	25	22	23	21	21
3D	22	24	21	19	25	24	25	25	22	23	22	22
4A	23	25	22	20	26	25	25	26	23	23	22	22
4B	23	25	22	20	27	26	26	27	24	24	23	23
5	24	26	23	21	30	27	27	28	25	24	23	23
6	25	27	24	22	31	28	28	29	26	25	24	24
7	33	40	29	27	36	32	32	36	30	28	26	27
8	36	45	31	29	38	34	34	38	32	29	27	28
9	37	46	32	30	39	36	35	40	33	30	28	29
10	38	47	33	31	39	37	35	41	33	31	29	29
11	40	48	34	32	40	38	36	42	34	32	30	30
12	49	64	40	38	53	52	46	65	44	39	36	37
13	50	64	40	38	54	53	47	66	45	40	37	38
14	59	78	45	43	64	62	55	81	53	45	41	43
15	60	80	46	44	69	69	58	91	56	48	44	46
16	61	82	47	45	74	76	61	101	59	51	47	49
17	63	86	49	47	84	90	67	121	65	57	53	55
18	64	88	50	48	86	92	68	122	66	58	54	56
19	66	90	50	48	88	94	70	124	68	58	54	56
20	67	91	51	49	89	95	71	125	69	59	55	57
21	67	92	52	50	89	95	71	126	69	60	56	58
22	68	93	53	51	90	96	72	127	70	61	57	59
23	72	100	56	54	94	99	76	133	74	63	59	61
24	76	104	58	56	100	106	80	144	78	66	62	64
25	78	106	60	56	102	108	82	146	78	66	62	64
26	79	108	61	57	102	109	83	147	79	67	63	65
27	80	109	62	58	103	110	84	148	80	68	64	66

TABLE TITLES BY VOLUME

Table Number	County of Clackmannan	County of Dumfries	County of Dunbarton	County of East Lothian	County of Fife	County of Inverness	County of Kincardine	County of Kirkcudbright	County of Lanark	County of Midlothian	Counties of Moray and Nairn
1	19	19	19	19	19	19	19	19	19	19	85
2	20	20	20	20	20	20	20	20	20	20	86
3A	20	21	21	21	21	21	20	21	21	21	86
3B	21	23	23	22	23	23	21	22	23	22	86
3C	21	24	24	22	25	25	21	23	25	22	87
3D	22	24	24	23	25	26	21	23	25	23	87
4A	22	25	24	23	26	26	22	23	26	23	87
4B	23	26	25	24	28	27	23	24	27	24	88
5	23	27	26	25	30	28	23	24	29	25	88
6	24	28	27	26	31	29	24	25	30	26	89
7	27	32	32	30	39	33	27	28	36	30	90
8	28	34	34	31	42	35	28	29	38	31	91
9	29	35	36	32	44	36	29	30	40	32	92
10	29	36	37	33	45	37	29	31	41	33	93
11	30	38	38	34	46	38	30	32	42	34	94
12	37	51	55	43	70	50	38	40	63	44	98
13	38	52	56	44	71	51	39	41	64	45	99
14	43	62	66	51	88	59	45	47	77	53	101
15	47	67	75	54	97	64	48	50	87	57	104
16	51	72	84	57	106	69	51	53	97	61	107
17	59	82	102	63	124	79	57	59	117	69	113
18	60	84	104	64	126	80	58	60	118	70	114
19	62	86	106	64	128	82	58	62	120	72	114
20	63	87	107	65	129	83	59	63	121	73	115
21	63	87	107	66	129	84	60	63	121	73	116
22	64	88	108	66	130	85	61	64	122	74	116
23	66	92	111	69	136	88	63	67	126	77	117
24	70	98	120	72	146	94	66	70	136	82	120
25	70	100	122	72	148	96	66	70	138	82	120
26	71	101	123	73	149	97	67	71	140	83	121
27	72	102	124	74	150	98	68	72	141	84	122

SCOTLAND

Table Number	County of Orkney	County of Peebles	Counties of Perth and Kinross	County of Renfrew	County of Ross and Cromarty	County of Roxburgh	County of Selkirk	County of Stirling	County of Sutherland	County of West Lothian	County of Wigtown	County of Zetland
1	19	19	19	111	19	19	19	19	19	19	19	19	19
2	20	20	20	112	20	20	20	20	20	20	20	20	20
3A	21	21	21	112	21	21	21	21	21	20	21	21	21
3B	22	22	23	112	23	22	22	22	23	21	22	22	22
3C	23	22	25	113	24	23	23	22	24	21	22	23	23
3D	23	23	25	113	24	23	23	23	24	22	23	23	23
4A	24	23	26	113	24	24	23	23	24	22	23	23	23
4B	25	24	27	114	25	25	24	24	25	23	24	24	24
5	26	24	29	114	26	26	25	24	26	23	25	24	24
6	27	25	30	115	27	27	26	25	27	24	26	25	25
7	30	27	35	116	32	31	29	27	32	26	30	27	29
8	31	28	37	117	34	33	30	28	34	27	31	28	30
9	32	29	38	118	36	34	31	29	36	28	32	29	31
10	33	29	39	119	37	35	31	29	37	29	33	29	31
11	34	30	40	120	38	36	32	30	38	30	34	30	32
12	43	37	54	124	56	46	40	37	54	36	44	37	41
13	44	38	55	125	57	48	41	38	55	37	45	38	42
14	51	43	65	127	67	55	46	42	65	41	52	43	49
15	54	46	70	130	77	58	50	46	72	44	57	46	52
16	57	49	75	133	87	61	54	50	79	47	62	49	55
17	63	55	85	139	107	67	62	58	93	53	72	55	61
18	64	56	86	140	108	68	64	60	94	54	74	56	62
19	64	56	88	140	110	70	66	62	96	54	76	56	62
20	65	57	89	141	111	70	67	63	97	55	77	57	63
21	66	58	89	142	112	71	67	63	98	56	77	58	64
22	67	59	90	142	113	72	68	64	99	57	78	59	65
23	70	61	94	143	116	76	70	66	102	59	81	61	68
24	74	64	100	146	126	80	74	70	110	62	86	64	72
25	74	64	102	146	128	80	74	70	112	62	86	64	72
26	75	65	103	147	129	81	75	71	113	63	87	65	73
27	76	66	104	148	130	82	76	72	114	64	88	66	74

TABLE TITLES BY VOLUME

VOLUME 2. USUAL RESIDENCE.[1]

TABLES

Number	Title	Page
1.	Comparison of enumerated and resident populations.[2]	1
2.	Visitors to Scotland (persons enumerated in but residing outside Scotland) by country of residence.[3]	12
3.	Visitors to Scotland (persons enumerated in but residing outside Scotland) by groups of country of residence.[4]	14
4.	Visitors to Scotland by age, by marital condition.	16
5.	Vistors to Scotland, by groups of country of residence, by age and marital condition.	16
6.	Visitors to Scotland by age.[5]	17

---

[1] The figures on resident population are based on the statements of usual residence in the census schedules. Unless otherwise noted, the area for which statistics are given is all of Scotland. No sampling is involved.

[2] Scotland, regional divisions and subdivisions, burghal/landward aggregates, cities, counties, large burghs, small burghs, districts of county, conurbation center, new towns.

[3] Scotland, regional divisions and subdivisions.

[4] Cities, counties, large burghs, small burghs with populations of 10,000 or more, county remainders, new towns.

[5] Scotland, regional divisions and subdivisions, burghal/landward aggregates.

SCOTLAND

VOLUME 3. AGE, MARITAL CONDITION, AND GENERAL TABLES.[1]

TABLES

Number	Title	Page
	Population, General	
1.	Population by sex, with intercensal changes, 1801-1961.	1
2.	Estimated population by sex as at the middle of each year, 1801-1961.	1
3.	Population at selected census, 1801-1961, and 1961 acreage.[2]	2
4.	Proportional distribution of population at selected censuses, 1801-1961, and 1961 acreage.[2]	3
5.	Average annual rates of change of population between selected censuses, 1801-1961.[2]	4
6.	Number of areas by type and size of population.	5
7A.	Census populations 1951-1961: intercensal changes: persons resident per acre, 1961.[3]	6
7B.	Population by area.[4] Part 1. In order of population size. Part 2. In alphabetical order.	12
7C.	Population, 1951 and 1961, and 1961 acreage.[5]	14
7D.	Population by sex.[6]	30
7E.	Population, 1951 and 1961.[7]	62

---

[1] Unless otherwise noted, the area for which statistics are given is all of Scotland. No sampling is involved.

[2] Scotland, regional divisions and subdivisions, administrative aggregates [the four counties of cities; 20 large burghs; 174 small burghs; and 33 landward areas], cities, counties.

[3] Scotland, regional divisions and subdivisions, burghal/landward aggregates, cities, counties, large burghs, small burghs, districts of county.

[4] Cities, large burghs, small burghs.

[5] Civil parishes.

[6] Registration districts (districts for registering births, deaths, and marriages).

[7] Inhabited islands.

7F.	Population by sex, electorate, electors per 100 population.[8]	63

Age and Marital Condition

8.	Age (single years) by marital condition.	65
9.	Age: numbers at successive censuses, 1841-1961.	67
10.	Age: proportions at successive censuses, 1841-1961, per 100,000 persons.	68
11.	Age: proportions at successive censuses, 1841-1961, per 100,000 of each sex.	69
12.	Age by marital condition at successive censuses, 1851-1961.	70
13.	Marital condition: proportions at successive censuses, 1851-1961, per 1,000 in each sex-age-group.	72
14.	Marital condition: proportions and sex ratios by age, 1921-1961.	74
15.	Population by age and sex.[9]	75
16.	Age by marital condition.[9]	78
17.	Age: proportions per 1,000 of each sex, 1951 and 1961.[9]	82
18.	Marital condition: proportions per 1,000 in each sex-age-group.[9]	84
19.	Sex, age, and marital condition indices.[10]	86

Non-private Households

20.	Population enumerated in private and non-private households.[11]	98
21.	Institutions: number, total population and inmates.[11]	100
22.	Hotels and boarding houses: number, rooms and population.[11]	105

---

[8] Parliamentary constituencies.

[9] Scotland, regional divisions and subdivisions, burghal/landward aggregates.

[10] Scotland, regional divisions and subdivisions, burghal/landward aggregates, cities, counties, large burghs, small burghs, districts of county, conurbation center, new towns.

[11] Scotland, regional divisions and subdivisions, burghal/landward aggregates, cities, counties.

SCOTLAND

23.	Population enumerated in and out with private households by sex and age.	106
24.	Institutions: age and marital condition of inmates.	107
25.	Children's homes: age of inmates.	109
26.	Hotels and boarding houses: age and marital conditions of guests.	109

VOLUME 4. HOUSING AND HOUSEHOLDS. PART II. HOUSEHOLD COMPOSITION TABLES.[1]

TABLES

Number	Title	Page
1.	Dwellings: rooms by household spaces (de jure) and households (de facto).[2]	1
2.	Households: persons (de jure and de facto) by rooms by sharing of dwelling.	2
3.	[See Footnote 3.]	
4.	Households: socio-economic group of chief economic supporter by tenure by persons by rooms.	6
5.	Households: household type by persons.	27
6.	Households: household type by persons by children.	28
7.	Households: household type by persons by rooms.	32
8.	Households: socio-economic group of chief economic supporter by household type by persons.	42

---

[1] Non-private households are excluded from these tables. Such households include hotels, boarding houses, prisons, etc. where the number of boarders, inmates, patients, or staff is five or more. Unless otherwise stated the data are based on a 10 percent sample, and the area for which statistics are given is all of Scotland.

[2] One percent sample.

[3] In common with the tables in other census reports, the design and numbering of the tables in this volume correspond to those adopted for similar tables published for England and Wales. This procedure was adopted to facilitate comparisons between English/Welsh and Scottish figures, and to enable Great Britain figures to be derived. A Scottish version of Table 3 was not produced, however, because the special post-census collection of data concerning households which were entirely absent on census night was effected in England and Wales only.

9.	Households and families by type.[4]	63
10.	Households: family type of chief economic supporter by his economic activity and household type.	76
11.	Households containing persons of pensionable age: household type. Persons of pensionable age in households: household type by sex, age, marital condition and economic activity.	78
12.	One- and two-person households containing persons of pensionable age: household type by rooms by sharing of dwelling.[5]	79
13.	One- and two-person households containing persons of pensionable age: sex/condition combination.[6]	95
14.	Husbands' and wives' ages in combination by socio-economic group of husband.	96
15.	Households: persons by earners by children.	103
16.	Households: persons by earners and children.[5]	105
17.	Households: socio-economic group of chief economic supporter by persons by earners and children.	109
18.	Households: socio-economic group of chief economic supporter by persons, earners and children.[5]	114
19.	Households: industry of chief economic supporter by persons and earners.	126
20.	Households: household type by earners by persons.	127
21.	Households: household type by persons and earners.[5]	130
22.	Households: persons by household type, earners and children.[7]	147
23.	Households containing visitors: persons in household by visitors. Visitors: persons in household by type of visitor. Visitor families: persons in household by family type.	162

---

[4] Scotland, regional divisions and subdivisions, burghal/landward aggregates, cities, counties, large burghs, small burghs, districts of county, conurbation center, new towns.

[5] Scotland, regional divisions and subdivisions, burghal/landward aggregates.

[6] Scotland, regional divisions and subdivisions, burghal/landward aggregates (100 percent sample).

[7] Scotland, cities, counties, large burghs, small burghs with populations of 10,000 or more, county remainders, new towns.

24.	Households containing domestic servants: domestic servants by persons in household. Domestic servants: sex and marital condition by persons in household.	162
25.	Households containing domestic servants and domestic servants' families: household type and family type.	163
26.	Households containing children, persons of pensionable age and domestic servants: household type.[5]	164
27.	Persons in households: household type by whether in families and economic activity.[5]	172
28.	Households with chief economic supporter born outside the British Isles: household type by country of birth.	180
29.	Households with chief economic supporter born outside the British Isles: country of birth by persons by rooms by sharing of dwelling.	181
30.	Families: household type by persons by family type.	185
31.	Families: household type by children by family type.	189
32.	Families: socio-economic group of head by children by family type.	193
33.	Families: household type by socio-economic group of head.	197
34.	Families: socio-economic group of head by persons by earners.	198
35.	Families: socio-economic group of head by family type by earners.	200
36.	Families: persons by family type.[5]	205
37.	Families: household type by persons and children.[5]	209
38.	Families: persons by socio-economic group of head.[5]	217
39.	Families: children's ages in combinations.	221
40.	Families and children: ages of parents by ages of children.	222
41.	Wives and mothers: economic activity and hours worked by number and age of children in family.	223
42.	Married couples: socio-economic group of husband by economic activity of husband and wife.	224
43.	Married couples: husband's socio-economic group by wife's economic activity by her age.	225
44.	Married couples: wife's economic activity by her age by number of children.	227
45.	Married couples with children in specified age groups: wife's economic activity by her age.	227

46.	Married couples with wife economically active: socio-economic group of husband and wife in combination.	228
AB.	Density (persons per room) by size and type of household and family type of chief economic supporter and comparison with hypothetical standards.	229
AD.	Households by family type of chief economic supporter and rooms occupied and comparison with hypothetical standards.	234

VOLUME 5. BIRTHPLACE AND NATIONALITY.[1]

TABLES

Number	Title	Page
1.	Birthplaces and nationalities of the whole population.[2]	1
2.	Birthplaces, nationalities and citizenships of residents of Scotland born outside the British Isles.[3]	3
3.	Nationalities and citizenships of residents of Scotland born outside the British Isles or with birthplace not stated.[3]	17
4A.	Birthplaces of the whole population: numbers.[4]	22
4B.	Birthplaces of the whole population: proportions per 10,000 of each sex.[4]	26
4C.	Birthplaces of persons born in Scotland.[5]	30
5.	Commonwealth citizens and citizens of the Irish Republic resident in but born outside Scotland: age and marital condition by country of birth.	35
6.	Aliens resident in Scotland: age and marital condition by country of nationality.	38

---

[1] Unless otherwise noted, the area for which statistics are given is all of Scotland. No sampling is involved.

[2] Scotland, regional divisions and subdivisions.

[3] Scotland, regional divisions and subdivisions, cities, counties with 2,000 or more persons born outside the British Isles.

[4] Scotland, regional divisions and subdivisions, cities, counties, large burghs, small burghs with populations of 10,000 or more, county remainders, new towns.

[5] Scotland, regional divisions and subdivisions, cities, counties.

Number	Title	Page
7.	Birthplaces of the population at selected censuses, 1851-1961.	39
8.	Birthplaces of the population at selected censuses, 1851-1961 (proportions per 100,000 persons).	39
9.	Scottish-born persons enumerated at selected censuses, 1851-1961.[6]	39

VOLUME 6. OCCUPATION, INDUSTRY, AND WORKPLACE. PART I. OCCUPATION TABLES.[1]

TABLES

Number	Title	Page
1.	Occupation by status.	2
2.	Occupation and status by marital condition and age.	8
3.	Occupation and status by marital condition by age.	18
4.	Self-employed without employees: occupation by marital condition and age.	42
5.	Self-employed with employees (large establishments): occupation by marital condition and age.	44
6.	Self-employed with employees (small establishments): occupation by marital condition and age.	46
7.	Managers (large establishments): occupation by marital condition and age.	48
8.	Managers (small establishments): occupation by marital condition and age.	49
9.	Foremen and supervisors (manual): occupation by marital condition and age.	50
10.	Foremen and supervisors (non-manual): occupation by marital condition and age.	52
11.	Apprentices, articled clerks and formal trainees: occupation by marital condition and age.	53
12.	Other employees (including professional): occupation by marital condition and age.	55
13.	Persons out of employment (including sick): occupation by marital condition and age.	60

---

[6] Countries of the British Isles.

[1] Ten percent sample. Unless otherwise noted, the area for which statistics are given is all of Scotland.

TABLE TITLES BY VOLUME

14.	Part-time workers: occupation by marital condition and age.	62
15.	Family workers: occupation by marital condition and age.	64
16.	Institution inmates by occupation.	66
17.	Retired persons: former occupation by marital condition and age.	67
18.	Percentage retired by age (single years) by occupation and socio-economic group.	74
19.	Socio-economic group by marital condition by age.	112
20.	Social class[2] by marital condition by age.	120
21.	Males in part-time employment: occupation in former full-time employment by marital condition and age.	121
22.	Males in part-time employment: present occupation by occupation in former full-time employment.	122
23.	Males in part-time employment: present socio-economic group in former full-time employment.	124
24.	Persons in part-time employment: hours worked by occupation.	125
25.	Persons in part-time employment: hours worked by marital condition by age.	127
26.	Occupation and status.[3]	130
27.	Socio-economic group and social class.[2,4]	154
28.	Socio-economic group (numbers and proportions) of economically active males.[3]	160
29.	Socio-economic group (numbers and proportions) of economically inactive males (stating an occupation).[3]	162

---

[2] The <u>social class</u> designation is determined by arranging the large number of unit groups of the occupational classification into a small number of categories, each of which is supposed to be homogeneous with respect to the general standing of the occupations concerned within the community. The categories are: I. Professional occupations; II. Intermediate occupations; III. Skilled occupations; IV. Partly skilled occupations; V. Unskilled occupations.

[3] Scotland, regional divisions and subdivisions, national and regional burghal/landward aggregates.

[4] Regional divisions and subdivisions, burghal/landward aggregates.

SCOTLAND

30.	Residents born outside Scotland: occupation by country of birth or nationality.	164
31.	Residents born outside Scotland: occupation by whether born in the United Kingdom and whether Commonwealth citizen or alien.[5]	169
A.	Proportions per 10,000 economically active of each sex by occupation.[3]	176
B.	Age proportions and proportion of females married per 1,000 of each sex economically active by occupation.	194

VOLUME 6. OCCUPATION, INDUSTRY, AND WORKPLACE. PART II. INDUSTRY TABLES.[6]

TABLES

Number	Title	Page
1.	Industry (full list) and status.	2
2.	Industry by status.	4
3.	Industry and status by age.	10
4.	Employment status by economic position by age.	16
5.	Industry by socio-economic group.	20
6.	Industry by salary/wage earner group (employed persons).	28
7.	Industry by occupation.	32
8.	Industry and status.[7]	98
8A.	Industry and status (persons working in Scotland enumerated anywhere in Great Britain).[8]	116
9.	Males in part-time employment: present industry by industry in former full-time employment.	118
10.	Males in part-time employment: present status by status in former full-time employment.	119

---

[5] Regional divisions and subdivisions.

[6] Ten percent sample. Unless otherwise noted, the area for which statistics are given is all of Scotland.

[7] Regional divisions and subdivisions, national and regional burghal/landward aggregates, England, Wales, Northern Ireland.

[8] Great Britain, England and Wales.

11.	Residents born outside Scotland: industry by country of birth and nationality.	120
12.	Residents born outside Scotland: industry by whether born in the United Kingdom and whether Commonwealth citizen or alien.[9]	132
13.	Industry by status and occupation by age.	135
14.	Industry by status and occupation.[10]	178
A.	Proportion per 10,000 persons in employment by industry.[11]	224
B.	Age proportions per 1,000 of each sex and proportions of females married by industry.	234

VOLUME 6. OCCUPATION, INDUSTRY, AND WORKPLACE. PART III. WORKPLACE TABLES.[1]

TABLES

Number	Title	Page
1.	Population in employment by areas of residence and workplace.	2
2.	Comparison of resident and day population.	18
3A.	Areas of residence and workplace in combination.[2]	22

---

[9] Regional divisions and subdivisions.

[10] Scotland, regional divisions and subdivisions.

[11] Scotland, regional divisions and subdivisions, national and regional burghal/landward aggregates. England, Wales, Northern Ireland.

[1] Ten percent sample. This volume deals with residence-workplace commuting (workplace movement) between local authority areas, the conurbation center, and the new towns. Movement within these areas is not included. Persons who work in Scotland and reside elsewhere, and persons who live in Scotland and work elsewhere, are excluded also. Unless otherwise noted, the areas for which statistics are given are Scotland, regional divisions and subdivisions, burghal/landward aggregates, cities, counties, large burghs, small burghs, districts of county, conurbation center, new towns.

[2] Local authority areas.

SCOTLAND

3B.	Areas of residence and workplace in combination.[3]	56
3C.	Areas of residence and workplace in combination.[4]	58
4.	Areas of workplace and residence by socio-economic group, occupation and industry.[5]	59
5.	Workplace movement by socio-economic group, occupation and industry.[6]	60

VOLUME 7. GAELIC.[7]

TABLES

Number	Title	Page
1.	Persons speaking Gaelic (aged 3 and over), 1961. Numbers.[8]	1
2.	Persons speaking Gaelic (aged 3 and over), 1961. Numbers and percentage of total population.[9]	9
3.	Persons speaking Gaelic (aged 3 and over). Numbers and intercensal changes, 1951-1961.[9]	10
4.	Persons speaking Gaelic (aged 3 and over), 1961. Numbers 1961, and percentage of total population 1891-1961.[10]	11

---

[3] Conurbation, conurbation center, conurbation remainder.

[4] New towns.

[5] Conurbation.

[6] Local authority areas where sample figure of inward or outward movement is 1,000 or more.

[7] Of all persons aged 3 years and over information was gathered as to whether they spoke Gaelic or Gaelic and English. Unless otherwise noted, the statistics given are for all of Scotland. No sampling is involved.

[8] Scotland, cities, counties, large burghs, small burghs (aggregates), landward areas; wards in cities; burghs, districts of county and electoral divisions in counties of Argyll, Bute, Caithness, Inverness, Perth, Ross and Cromarty, Sutherland.

[9] Scotland, cities, counties.

[10] Counties of Argyll, Inverness, Perth (highland district only), Ross and Cromarty, Sutherland: burghs and civil parishes.

TABLE TITLES BY VOLUME

5.	Persons speaking Gaelic (aged 3 and over), 1951 and 1961. Numbers and intercensal changes.[11]	14
6.	Persons speaking Gaelic (aged 3 and over) by age, sex, 1961. Numbers.[12]	16
7.	Persons speaking Gaelic (aged 3 and over) by ages, 1961. Numbers and percentages.	27
8.	Persons speaking Gaelic (aged 3 and over) by ages, 1921-1961. Numbers.	27
9.	Persons speaking Gaelic (aged 3 and over) by ages, 1951 and 1961. Numbers.[13]	28

VOLUME 8. INTERNAL MIGRATION.[1]

TABLES

Number	Title	Page
1.	Resident population by sex and duration of residence.[2]	1

---

[11] Counties of Argyll, Inverness, Perth (highland district only), Ross and Cromarty, Sutherland: burghs and civil parishes with Gaelic-speaking population of 100 or more.

[12] Scotland, cities, counties, large burghs, areas in highland counties with 10,000 population or more, county remainders.

[13] Counties of Argyll, Inverness, Perth, Ross and Cromarty, Sutherland.

[1] Ten percent sample. The data in this volume come from a question on usual address a year before the census date. A *migrant* is anyone whose usual address differed between the two dates. An *immigrant* to an area was resident there at the time of the census but resident elsewhere a year before. An *emigrant* was resident in the area a year before but resident elsewhere in Scotland at the time of the census. Persons leaving Scotland are not included. Type and distance of move classifies migrants by the distance between area of residence at census and former area of residence, and also by whether the move was between different burghal areas, different landward areas, or between burghal and landward areas. Unless otherwise noted, the area for which statistics are given is all of Scotland.

[2] Scotland, regional divisions and subdivisions, burghal/landward aggregates, cities, counties, large burghs, small burghs, districts of county, conurbation center, new towns.

SCOTLAND

2. Population by sex and duration of residence by age, family status and socio-economic group.   13
3. Economically active resident population by sex by duration of residence by occupation and industry.   14
4. Migrants: numbers and proportions.[2]   16
5. Immigrants by area of former usual residence and sex.[3]   24
6. Immigrants by area of former usual residence and sex.[4]   25
7. Immigrants and emigrants by area of usual residence at census and former usual residence.[5]   26
8. Immigrants and emigrants by families.[5]   38
9. Immigrants by sex, age, type and distance of move.[6]   41
10. Emigrants by sex, age, type and distance of move.[7]   47
11. Immigrants in employment moving to districts of county outside conurbation by sex and workplace.[8]   52
12. Migrants by sex, age, family status, type and distance of move.   53
13. Migrants by sex, family status and type of move.[7]   61
14. Wholly moving private households[9] by socio-economic group of chief economic supporter, tenure and density of occupation.[10]   64

---

[3] Regional divisions and subdivisions.

[4] Burghal/landward aggregates.

[5] Cities, counties, large burghs, small burghs, districts of county and new towns with 50 or more immigrants or emigrants in sample.

[6] Scotland, regional divisions and subdivisions, burghal/landward aggregates.

[7] Regional divisions and subdivisions, burghal/landward aggregates.

[8] Scotland, regional divisions.

[9] **Wholly moving private households** are those households enumerated at their usual residence, all of whose members present on census night (aged 1 or over) had changed their usual residence in the year before the census date. All such members need not have moved from the same area of former usual residence.

[10] Scotland, regional divisions and subdivisions, burghal/landward aggregates, cities, counties, large burghs, small burghs,

TABLE TITLES BY VOLUME

15. Wholly moving families[11] enumerated at usual residence by type of move, status of family head,[12] size of family and persons per room.   73

16. Migrants by sex, type and distance of move, occupation, industry and socio-economic group.[6]   75

17. Immigrants from outside Scotland by sex, age, family status and former country of residence.[6]   123

18. Aliens (including stateless) resident in Scotland, born outside the United Kingdom by former country of residence and country of nationality.   127

VOLUME 9. TERMINAL EDUCATION AGE.[1]

TABLES

Number	Title	Page
1.	Population aged 15 and over in 9 age sections classified by 16 terminal education age groups.	2
2.	Population aged 15 and over in 9 age sections classified by 7 terminal education age groups.[2]	4

---

districts of county, conurbation center, or new towns with 50 or more wholly moving households in sample.

[11]<u>Wholly moving families</u> are those families enumerated at their usual residence, all of whose members present on census night (aged 1 or over) had moved from the same area of former usual residence.

[12]<u>Status of family head</u> is determined by dividing families into two types: (1) married couples with their never-married children (if any), in which case the husband is the head; (2) lone parents with never-married children, in which case the lone parent is the head.

[1]Ten percent sample. This volume gives figures on the age at which full-time education at school, college, or university ceased. Respondents were asked for the following information: "For all persons aged 15 and over not now receiving a full-time education at school, college, university, etc., write the age at which such education ended." Students who were enrolled in full-time study but were also employed were not counted in this question. Unless otherwise noted, the area for which statistics are given is all of Scotland.

[2]Regional divisions and subdivisions, burghal/landward aggregates.

SCOTLAND

3. Population aged 15 and over by socio-economic group and economically active population by occupation, each classified by age and 6 terminal education age groups.   12

4. Population aged 25 and over by 7 terminal education age groups.[3]   22

VOLUME 10. FERTILITY.[1]

TABLES

Number	Title	Page
1(i).	Married women by age at census and size of family.	1
1(ii).	Widowed and divorced women by age at census and size of family.	2

---

[3] Cities, counties, new towns.

[1] The tables in this volume comprise all published statistics derived from the questions on the number of children born alive in marriage and whether any of these children were born in the year before the census. These questions were put to all married, widowed, and divorced women. The <u>duration of present marriage</u> derives from the difference between the marriage and the census dates; <u>age at present marriage</u> derives from the difference between the age at the census and the year of marriage. The <u>duration of terminated first or only marriage</u> derives from information on the date of the marriage and its termination. <u>Married women</u> are women married at the time of the census, regardless of whether it is their first or a later marriage. <u>Widowed and divorced women</u> are women who were widowed or divorced at the time of the census irrespective of the number of times they have been married. <u>Women married once only</u> are women married under 45 whose only marriage was still in existence at census date. <u>Women with uninterrupted first marriage</u> are women: (1) whose only marriage took place before they were 45 and was still in existence at the census date, or (2) whose first or only marriage took place before their 45th birthday and lasted until they were 45 years old, irrespective of their marital status at census date. <u>Women married once only and enumerated with their husbands</u> are women married under the age of 45 whose only marriage was still in existence at census date and who were included on the same census schedule as their husbands. <u>Remarried women</u> are women in their second or a later marriage at census date. These tables relate to the legitimate fertility of women and thus exclude any children born to these women before marriage or between marriages. Unless otherwise noted, the area for which statistics are given is all of Scotland. No sampling is involved.

TABLE TITLES BY VOLUME

2(i).	Married women: size of family by duration of current marriage and age at current marriage.	3
2(ii).	Current fertility of married women: age at current marriage by duration of current marriage and size of family.	6
3.	Women with uninterrupted first marriage: age at census by duration of marriage: number of women, number of children, mean family size, number of women infertile and proportion infertile.	8
4.	Women with uninterrupted first marriage: size of family by age at census.	18
5(i).	Women with uninterrupted first marriage: size of family by duration of marriage and age at census.	19
5(ii).	Current fertility of women with uninterrupted first marriage: by age at census, duration of marriage and size of family.	24
6(i).	Women with uninterrupted first marriage: size of family by duration of marriage and age at marriage.	25
6(ii).	Current fertility of women with uninterrupted first marriage: age at marriage by duration of marriage and size of family.	28
7.	Women with uninterrupted first marriage: size of family by year of marriage and age at marriage.	29
8(i).	Remarried women (excluding women with uninterrupted first marriage): size of family by age at current marriage and duration of current marriage.	36
8(i). Supplement	All remarried women with current marriage age 45 and over: size of family by age at current marriage and duration of current marriage.	39
8(ii).	Current fertility of remarried women (excluding women with uninterrupted first marriage): age at current marriage, duration of current marriage and size of family.	40
9(i).	Remarried women (excluding women with uninterrupted first marriage): size of family by age at first marriage and time since first marriage.	42
9(ii).	Current fertility of remarried women (excluding women with uninterrupted first marriage) by age at first marriage, time since first marriage and size of family.	45

SCOTLAND

10(i). Women married once only and enumerated with their husbands: size of family by age at marriage and duration of marriage.[2] — 47

10(ii). Current fertility of women married once only and enumerated with their husbands: by age at marriage, duration of marriage and size of family.[2] — 50

11. Women married once only and enumerated with their husbands: size of family by husband's age at census.[2] — 51

12(i). Ratios of the fertility of women with uninterrupted first marriage to that of all married women. — 52

12(ii). Ratios of the fertility of women married once only and enumerated with their husbands to that of women with uninterrupted first marriage. — 53

12(iii). Ratios of the fertility of remarried women (excluding women with uninterrupted first marriage) classified by current marriage to that of women with uninterrupted first marriage. — 54

12(iv). Ratios of the fertility of remarried women (excluding women with uninterrupted first marriage) classified by first marriage to that of women with uninterrupted first marriage. — 55

13. Women with uninterrupted first marriage: fertility by age at marriage and duration of marriage.[3] — 56

14. Women married once only and enumerated with their husbands: fertility by age at marriage, duration of marriage and socio-economic group of husband.[2] — 86

15. Women married once only and enumerated with their husbands: fertility by age at marriage, duration of marriage and difference between ages of husband and wife.[2] — 124

16. Women married once only and enumerated with their husbands: fertility by age at marriage, duration of marriage and industry group of husband.[2] — 138

17. Women married once only and enumerated with their husbands: fertility by age at marriage, duration of marriage and occupation order of husband.[2] — 164

18. Women married once only and enumerated with their husbands: fertility by age at marriage, duration

---

[2] Ten percent sample.

[3] Regional divisions and subdivisions, burghal/landward aggregates.

	of marriage and terminal education age of husband and wife.[2]	218
19.	Women married once only and economically active: fertility by age at marriage, duration of marriage and occupation group.[2,4]	250
20.	Women with uninterrupted first marriage: fertility by age at marriage, duration of marriage and citizenship and country of birth.	262
Appendix A.	Women married once only: age at census by year of marriage.	286

---

[4] The <u>occupation groups</u> are a selection of those occupation orders with large numbers of women, namely (1) textile and clothing workers, (2) sales workers, (3) professional, technical workers, artists, (4) service, sport and recreation workers, (5) clerical workers.

SPAIN

MAIN ENTRIES AND VOLUME NOTES:

Spain. Instituto Nacional de Estadística.
Censo de la población y de las viviendas de España de 1960. Nomenclator de las ciudades, villas, lugares, aldeas, y demas entidades de población. [Madrid ?] n.d.
51 fascicles. 28cm.

Alava	Cuenca	Navarra	Toledo
Albacete	Gerona	Orense	Valencia
Alicante	Granada	Oviedo	Valladolid
Almeria	Guadalajara	Palencia	Vizcaya
Avila	Guipúzcoa	Palmas (Las)	Zamora
Badajoz	Huelva	Pontevedra	Zaragoza
Baleares	Huesca	Salamanca	[1] Plazas de
Barcelona	Jaén	Santa Cruz de	soberanía en
Burgos	León	Tenerife	el Norte
Cáceres	Lérida	Santander	de Africa
Cádiz	Logroño	Segovia	[2] Resumenes
Castellón	Lugo	Sevilla	generales y
Ciudad Real	Madrid	Soria	apéndice de
Córdoba	Málaga	Tarragona	variaciones
Coruña (La)	Murcia	Teruel	

Spain. Instituto Nacional de Estadística.
Censo de la población y de las viviendas de España, según la inscripción realizada el 31 de diciembre de 1960. [Madrid, 1962-1969]
5 v.   (64-38486)

Tomo I. Cifras generales de habitantes. [1962] xl, 404 p.

Tomo II. Cifras generales de viviendas. [1964] vii, 207 p. [Excluded.]

Tomo III. Población.[1]
  Volumen 1. Resumen nacional y provincias de Alava a La Coruña. [1969] xi, 16 fascicles.
  Volumen 2. Provincias de Cuenca a Oviedo. [1969] 18 fascicles.

---

[1] Data in this tome are based on a 25 percent sample.

Volumen 3. Provincias de Palencia a Zaragoza y municipios de Ceuta y Melilla. [1969] 18 fascicles.

Tomo IV. Classificaciones de las viviendas. [1969] iii, 51 fascicles. [Excluded.]

Tomos III y IV. Fasciculo No. 0. Resumen nacional. [1969] 2 fascicles.[2] [Excluded; duplicate data.]

Spain. Instituto Nacional de Estadística.
Censo de población y de las viviendas 1960. Población de derecho y hecho de los municipios de la nación. Madrid, 1962. 104 p. 22 cm.

ENGLISH TRANSLATION OF CENSUS AND VOLUME TITLES:

Census of population and housing of Spain 1960. Nomenclature for the cities, towns, villages, hamlets, and other population entities.

Census of population and housing of Spain, according to the enumeration of December 31, 1960.
Tome I. General figures on the inhabitants.

Tome II. General figures on housing.

Tome III. Population.
Volume 1. National summary and provinces Alava through Coruña.
Volume 2. Provinces Cuenca through Oviedo.
Volume 3. Provinces Palencia through Zaragoza and the municipalities of Ceuta and Melilla.

Tome IV. Housing.

Tomes III and IV. Fascicle No. 0. National summary.

Census of population and housing 1960. De jure and de facto population of the municipalities of the nation.

ENUMERATION DATE: December 31, 1960.

ENUMERATION BASIS: De facto; de jure.

---

[2] The national summary data of Tomes III and IV are available bound as one unit.

SPAIN

GLOSSARY

ADMINISTRATIVE, GEOGRAPHICAL, AND POLITICAL UNITS

Level	English Translation	Vernacular Term
National	The Spanish State	Estado español
Intermediate	Metropolitan province	Provincia metropolitana
	African provinces and territories	Provincias y plazas africanas
Local	Municipality	Municipio

Other concepts:

	Zones	Zonas
	Rural	Rural
	Intermediate	Intermedia
	Urban	Urbana

The census was taken in all territories under Spanish jurisdiction. The national territory consists of the _metropolitan provinces_ (the 50 provinces of the Iberian peninsula and the Balearic and Canary Islands). The _African provinces and territories_ are made up of the territories of North Africa, the provinces of the Equatorial Region, and the provinces of Ifni and Sahara.

The metropolitan provinces are divided into _municipalities_, which are autonomous from each other in their municipal spheres and are managed by municipal councils.

Provinces are divided into _zones_, which are classified, according to the size of the population nucleus within the zone, as follows: _rural zone_, fewer than 2,000; _intermediate zone_, 2,000 to 10,000; and _urban zone_, more than 10,000 inhabitants.

TABLE OF DEMOGRAPHIC CONCEPTS

The demographic concepts which appear in the following table correspond to the recommendations of the European Conference of Statisticians, unless otherwise noted. They are defined and discussed in the appendixes and in footnotes. Terms in brackets are not used in the table titles, but they are included here in order to give logical completeness to the various classifications in the table.

GLOSSARY

Appendix	English Table Term	Vernacular Table Term
CULTURAL, EDUCATIONAL, AND PERSONAL CHARACTERISTICS		
...	Level of education[1]	Estudios
	In progress	En curso
	Completed	Terminados
	Literacy[2]	Instrucción elemental
	Nationality[3]	Nacionalidad
	Place of birth[4]	Naturaleza
ECONOMIC CHARACTERISTICS		
A	Economic activity	Actividad económica
	Economically active	Económicamente activa
	Employed	Personas que ejercen una profesión
	[Unemployed]	...
	Economically inactive	Económicamente no activa
	Independent	Independientes
	Dependent	Dependientes
B	Employment status	Posición en la ocupación
C	[Industrial classification]	...
	Division(s)	Características de los establecimientos donde trabajan
D	[Occupational classification]	...
	Major group(s)	Grupos profesionales
FAMILY CHARACTERISTICS		
...	Fertility	Fecundidad
	Marital status	Estado civil

---

[1] Level of education (whether in progress or complete) refers to tabulations roughly corresponding to primary, secondary, or college/university in the United States.

[2] Literacy refers to tabulations for the literate and the illiterate, illiterate persons being defined as those over 10 years of age who cannot read and write a simple account of facts related to their everyday life.

[3] Nationality is the country of which a person is a citizen or subject.

[4] Place of birth is the municipality in which a Spaniard was born, or nation (by 1960 territorial definitions) in which an alien was born.

SPAIN

TABLE TITLES BY VOLUME

CENSUS OF POPULATION AND HOUSING OF SPAIN. NOMENCLATURE FOR THE CITIES, TOWNS, VILLAGES, HAMLETS, AND OTHER POPULATION ENTITIES. [50 fascicles; see listing on p. 348 above.]

Structure of the Fascicles

These fascicles are a compilation of tables, all of the same format, for the 50 provinces. The tables consist of a variety of data for (1) the municipalities and their subdivisions (called individual entities), and (2) summary figures for judicial districts and the kinds of individual entities within them.

(1) Municipalities and subdivisions (individual entities)

    Category of individual entity
    Area in square kilometers
    Altitude at the capital
    Population
      De jure
      De facto
    Data for clustered buildings
      Number of dwellings
      Number of hotels, etc.
      Number of households
        Private
        Institutional
      Normally resident population[1]
        Total
        Male
        Female
    Data for scattered buildings
      Number of dwellings
      Number of hotels, etc.
      Number of households
        Private
        Institutional
      Normally resident population[1]
        Total
        Male
        Female

---

[1] The normally resident population is a separate concept from the de jure and the de facto populations: it is a count of the normal inhabitants of the housing.

(2) <u>Summary by judicial districts</u>
    Entities with a nucleus
      Number of entities
      Data for clustered buildings
        Number of dwellings
        Number of households
        Normally resident population[1]
      Data for scattered buildings
        Number of dwellings
        Number of households
        Normally resident population[1]
    Entities without a nucleus
      Number of entities
      Number of dwellings
      Number of households
      Normally resident population[1]

    Number of municipalities
    Individual entities
      Total
      Cities
      Towns
      Villages
      Hamlets
      Groups of houses
      Other
    Population
      De jure
      De facto

CENSUS OF POPULATION AND HOUSING, ACCORDING TO THE ENUMERATION OF DECEMBER 31, 1960.

TOME I. GENERAL FIGURES ABOUT THE INHABITANTS.[2]

TABLES

Number	Title	Page
	Provincial Censuses	
1.	Alava	1
2.	Albacete	5
3.	Alicante	9

---

[2] The provincial census tables have figures for the <u>de jure</u> population by municipality; residents, absent and present, by sex; non-residents temporarily present; and the <u>de facto</u> population.

SPAIN

4.	Almeria	15
5.	Avila	19
6.	Badajoz	27
7.	Baleares	33
8.	Barcelona	37
9.	Burgos	47
10.	Cáceres	61
11.	Cádiz	69
12.	Castellón	72
13.	Ciudad Real	78
14.	Córdoba	82
15.	Coruña (La)	86
16.	Cuenca	90
17.	Gerona	99
18.	Granada	107
19.	Guadalajara	114
20.	Guipúzcoa	126
21.	Huelva	130
22.	Huesca	134
23.	Jaén	144
24.	León	148
25.	Lérida	156
26.	Logroña	165
27.	Lugo	172
28.	Madrid	176
29.	Málaga	183
30.	Murcia	187
31.	Navarra	190
32.	Orense	198
33.	Oviedo	202
34.	Palencia	206
35.	Palmas (Las)	214
36.	Pontevedra	217
37.	Salamanca	222

38.	Santa Cruz de Tenerife	234
39.	Santander	238
40.	Segovia	244
41.	Sevilla	251
42.	Soria	255
43.	Tarragona	265
44.	Teruel	272
45.	Toledo	281
46.	Valencia	288
47.	Valladolid	296
48.	Vizcaya	304
49.	Zamora	309
50.	Zaragoza	318
51.	African territories and provinces	327
	General Summaries	
1.	Provincial summary[3]	331
2.	Capital summary[3]	332
3.	General summary of the population of Spain[4]	333

TOME III. POPULATION.

Structure of the Tome

Tome III is divided into three volumes, which are composed of 52 fascicles as follows: Volume 1 contains Fascicles 0 and 1-15; Volume 2 contains Fascicles 16-33; Volume 3 contains Fascicles 34-51. Fascicle No. 0 is a National Summary; Nos. 1-50 have

---

[3]The provincial summary is arranged by judicial districts and has figures for the number of household schedules, private and institutional; the <u>de jure</u> population by sex; the <u>de facto</u> population by sex; and the number of dwellings by size group of the municipal population entities. The provincial and the capital summaries have figures for the number of household schedules; the <u>de jure</u> population; residents, absent and present, by sex; non-residents temporarily present, by sex; and the <u>de facto</u> population.

[4]The general summary has figures for the <u>de facto</u> and <u>de jure</u> populations of Spain and its territories.

SPAIN

tabulations for the 50 metropolitan provinces (see pp. 353-355 above for a listing of the provinces in the same numerical sequence as used for the fascicles); and No. 51 gives tabulations for the municipalities of Ceuta and Melilla, in Morocco.[1]

Fascicle No. 0 is arranged and paged uniquely. Fascicles Nos. 1-51 are identical as to table numbers and titles (except for tabulations for urban areas in the provinces and a somewhat modified treatment of the municipalities of Ceuta and Melilla),[2] but the page locations vary from fascicle to fascicle. The Tables list and page locations for Fascicle No. 0 are given immediately below, followed by the Tables list and Page Location Matrix which are to be used to locate desired data in Fascicles 1-51.

NATIONAL SUMMARY TABLES [FASCICLE 0]

Number	Title	Page
1.	Classification of the population by age group.	2
2.	Classification of the population by age, marital status, sex, and literacy.	
	2.1. Marital status, sex, and literacy.	8
	2.2. Age and sex.	9
3.	Classification of the population by place of birth, nationality, and marital status.	13
4.	Classification of the population by literacy and level of education.	
	4.1. Literacy and education in progress.	16
	4.2. Education completed.	19
5.	Classification of the population by economic activity.	
	5.1. Economically active population.	
	5.1.1. Employment status.	22
	5.1.2. Occupational major group.	25
	5.1.3. Industrial division.	28
	5.2. Economically inactive population.	
	5.2.1. Dependent or independent.	31
	5.2.2. Economically inactive population dependent upon employed persons, by the industrial division where their supporter works.	34
6.	Classification of women by their fertility.	37

---

[1] The municipalities of Ceuta and Melilla are each referred to as a "sovereign place," i.e., **plaza soberania** (see Tome I, pp. 354, 371).

[2] The arrangement of Fascicle No. 51 differs from Nos. 1-50 in that it has no title page and Table 2 has the following subdivisions: 2.1, data for Ceuta; 2.2, data for Melilla.

TABLE TITLES BY VOLUME

PROVINCIAL AND MUNICIPAL TABLES [FASCICLES 1-51]

Number	Title
1.	Classification of the population by age group.
2.	Classification of the population by age, marital status, sex, and literacy.
	2.1. Total for the province.
	2.2. Urban zone.
	2.3. Intermediate zone.
	2.4. Rural zone.
	2.5. Capital.
3.	Classification of the population by place of birth, nationality, and marital status.
4.	Classification of the population by literacy and level of education.
	4.1. Literacy and education in progress.
	4.2. Education completed.
5.	Classification of the population by economic activity.
	5.1. Economically active population.
	5.1.1. Employment status.
	5.1.2. Occupational major group.
	5.1.3. Industrial division.
	5.2. Economically inactive population.
	5.2.1. Dependent or independent.
	5.2.2. Economically inactive population dependent upon employed persons, by the industrial division where their supporter works.
6.	Classification of women by their fertility.

PAGE LOCATION MATRIX

Fascicle Number	1	2.1	2.2	2.3	2.4	2.5	2.6	2.7	2.8	2.9	2.10	3	4.1	4.2	5.1.1	5.1.2	5.1.3	5.2.1	5.2.2	6
1	2	4	6	8	10	12	-	-	-	-	-	14	15	16	17	18	19	20	21	22
2	2	4	6	8	10	12	-	-	-	-	-	14	15	16	17	18	19	20	21	22
3	2	6	8	10	12	14	16	18	-	-	-	20	22	24	26	28	30	32	34	36
4	2	4	6	8	10	12	-	-	-	-	-	14	15	16	17	18	19	20	21	22
5	2	4	6	8	10	12	-	-	-	-	-	14	15	16	17	18	19	20	21	22
6	2	6	8	10	12	14	-	-	-	-	-	16	18	20	22	24	26	28	30	32
7	2	6	8	10	12	14	-	-	-	-	-	16	18	20	22	24	26	28	30	32
8	2	8	10	12	14	16	18	20	22	24	26	28	31	34	37	40	43	46	49	52
9	2	4	6	8	10	12	-	-	-	-	-	14	15	16	17	18	19	20	21	22
10	2	4	6	8	10	12	-	-	-	-	-	14	15	16	17	18	19	20	21	22
11	2	8	10	12	14	16	18	20	22	24	-	26	29	32	35	38	41	44	47	50
12	2	6	8	10	12	14	-	-	-	-	-	16	18	20	22	24	26	28	30	32
13	2	6	8	10	12	14	16	-	-	-	-	18	20	22	24	26	28	30	32	34
14	2	8	10	12	14	16	-	-	-	-	-	18	21	24	27	30	33	36	39	42
15	2	8	10	12	14	16	18	20	-	-	-	22	25	28	31	34	37	40	43	46
16	2	4	6	8	10	12	-	-	-	-	-	14	15	16	17	18	19	20	21	22
17	2	4	6	8	10	12	-	-	-	-	-	14	15	16	17	18	19	20	21	22
18	2	6	8	10	12	14	-	-	-	-	-	16	18	20	22	24	26	28	30	32
19	2	4	6	8	10	12	-	-	-	-	-	14	15	16	17	18	19	20	21	22
20	2	6	8	10	12	14	-	-	-	-	-	16	18	20	22	24	26	28	30	32
21	2	6	8	10	12	14	-	-	-	-	-	16	18	20	22	24	26	28	30	32
22	2	4	6	8	10	12	-	-	-	-	-	14	15	16	17	18	19	20	21	22
23	2	8	10	12	14	16	18	-	-	-	-	20	23	26	29	32	35	38	41	44
24	2	4	6	8	10	12	-	-	-	-	-	14	15	16	17	18	19	20	21	22
25	2	4	6	8	10	12	-	-	-	-	-	14	15	16	17	18	19	20	21	22

Fascicle Number	1	2.1	2.2	2.3	2.4	2.5	2.6	2.7	2.8	2.9	2.10	3	4.1	4.2	5.1.1	5.1.2	5.1.3	5.2.1	5.2.2	6
26	2	4	6	8	10	12	-	-	-	-	-	14	15	16	17	18	19	20	21	22
27	2	6	8	10	12	14	-	-	-	-	-	16	18	20	22	24	26	28	30	32
28	2	4	6	8	10	12	-	-	-	-	-	14	15	16	17	18	19	20	21	22
29	2	6	8	10	12	14	-	-	-	-	-	16	18	20	22	24	26	28	30	32
30	2	8	10	12	14	16	18	20	-	-	-	22	25	28	31	34	37	40	43	46
31	2	4	6	8	10	12	-	-	-	-	-	14	15	16	17	18	19	20	21	22
32	2	4	6	8	10	12	-	-	-	-	-	14	15	16	17	18	19	20	21	22
33	2	8	10	12	14	16	18	20	22	-	-	24	27	30	33	36	39	42	45	48
34	2	4	6	8	10	12	-	-	-	-	-	14	15	16	17	18	19	20	21	22
35	2	6	8	10	12	14	-	-	-	-	-	16	18	20	22	24	26	28	30	32
36	2	8	10	12	14	16	18	-	-	-	-	20	23	26	29	32	35	38	41	44
37	2	4	6	8	10	12	-	-	-	-	-	14	15	16	17	18	19	20	21	22
38	2	6	8	10	12	14	16	-	-	-	-	18	20	22	24	26	28	30	32	34
39	2	4	6	8	10	12	-	-	-	-	-	14	15	16	17	18	19	20	21	22
40	2	4	6	8	10	12	-	-	-	-	-	14	15	16	17	18	19	20	21	22
41	2	8	10	12	14	16	-	-	-	-	-	18	21	24	27	30	33	36	39	42
42	2	4	6	8	10	12	-	-	-	-	-	14	15	16	17	18	19	20	21	22
43	2	4	6	8	10	12	-	-	-	-	-	14	15	16	17	18	19	20	21	22
44	2	4	6	8	10	12	-	-	-	-	-	14	15	16	17	18	19	20	21	22
45	2	4	6	8	10	12	-	-	-	-	-	14	15	16	17	18	19	20	21	22
46	2	8	10	12	14	16	-	-	-	-	-	18	21	24	27	30	33	36	39	42
47	2	4	6	8	10	12	-	-	-	-	-	14	15	16	17	18	19	20	21	22
48	2	6	8	10	12	14	16	-	-	-	-	18	20	22	24	26	28	30	32	34
49	2	4	6	8	10	12	-	-	-	-	-	14	15	16	17	18	19	20	21	22
50	2	4	6	8	10	12	-	-	-	-	-	14	15	16	17	18	19	20	21	22
51	2	3	5	-	-	-	-	-	-	-	-	7	7	8	8	9	9	10	10	11

SWEDEN

MAIN ENTRY AND VOLUME NOTES:[1]

Sweden. Statistiska Centralbyrån.
Folkräkningen den 1 november 1960. Stockholm, 1961-1965.
11 v. 24cm. (63-30152)

I. Folkmängd inom kommuner och församlingar efter kön, ålder, civilstånd m.m. 1961. vii, 214 p.

II. Folkmängd inom tätorter efter kön, ålder och civilstånd. 1961. vi, 137 p.

III. Folkmängd i hela riket och länen efter kön, ålder och civilstånd m.m. 1961. vii, 145 p.

IV. Folkmängd inom särskilda områden, utlänningar m.m. 1962. x, 147 p.

V. Indelningar, tätortsavgränsning, befolkningsutveckling m.m. 1962. xv, 118 p. [Excluded; no tables.]

VI. Förvärsarbetande befolkning efter näringsgren och yrkesställning m.m. inom kommuner, församlingar och tätorter. 1963. viii, 416 p.

VII. Förvärvsarbetande befolkning efter yrke samt hushåll efter hushållsstorlek m.m. inom kommuner och större tätorter. 1963. viii, 224 p.

VIII. Förvärvsarbetande dagbefolkning efter näringsgren samt utpendling och inpendling i kommuner och landsförsamlingar. 1963. viii, 242 p.

IX. Näringsgren, yrke, pendling, hushåll och utbildning i hela riket, länsvis m.m. 1964. xii, 244 p.

X. Huvudsysselsättning, åker- och skogsareal, personbilsinnehav samt nationalitet. 1964. x, 138 p.

XI. Urvalsbearbetning: familjer, inkomst, inrikes omflyttning och näringsgrensväxling. 1965. xii, 107 p.

---

[1] Each volume of the Swedish census has a separate title page in English, a summary in English of the introductory material in that volume, and a Swedish/English glossary. In addition, the table titles are given in both Swedish and English.

ENGLISH TRANSLATION OF CENSUS AND VOLUME TITLES:

Census of the population on November 1, 1960.[1]

- I. Population in communes and parishes by sex, age, and marital status, etc.
- II. Population in localities by sex, age, and marital status.
- III. Population in the whole country and in the counties, by sex, age, and marital status, etc.
- IV. Population in different divisions, aliens, etc.
- V. Divisions, demarcations of localities, development of the population, etc.
- VI. Economically active population by industry and status, etc., in communes, parishes, and localities.
- VII. Economically active population by occupation and households by size, etc., within communes and larger localities.

---

[1] This census was coordinated with the annual registration of the population, which took place on November 1, 1960, and it was done without regard to citizenship. In Sweden, each local parish keeps a *personakt*, or register, of its residents, which is forwarded from one parish to another on the change of residence of an individual. The parish office registers all local vital events, reports of which are forwarded regularly to the county, which in turn sends them to the Central Bureau of Statistics. The accuracy of both the parish and county registers is checked annually by the *mantalsskrivning*, a procedure requiring the head of every household to complete a form giving particulars on his household and its members. For a formal census, the *mantalsskrivning* is supplemented by a more detailed census questionnaire. (See E. v. Hofsten, "Population Registers and Computers: New Possibilities for the Production of Demographic Data," *International Statistical Institute Review* 34:2 (1966): 186-194, *passim*.) The census is taken entirely by self-enumeration. Volumes I-IV contain the results obtained from the annual *mantalsskrivning*. Volume V, not included here, describes and analyzes the data found in Volumes I-IV, and compares them with data from previous censuses. Volumes VI-X contain information from the special census questionnaire used with the *mantalsskrivning*. Due to corrections and/or the exclusion of data at this second stage, minor differences from the first four volumes appear in figures for a few communes, particularly in data on sex, age, marital status, year of marriage, and the demarcation of localities. There is considerable divergence between population figures for individual localities in the second stage and those figures in Volume II, resulting primarily from different definitions of "household."

SWEDEN

VIII. Economically active day-population by industry and outgoing and ingoing commuting in communes and rural parishes.

IX. Industry, occupation, commuting, households, and education in the whole country, by county, etc.

X. Main occupation, arable and forest acreage, private car owners, and nationality.

XI. Sample surveys: families, income, internal migration, and change of industry.[2]

ENUMERATION DATE: November 1, 1960.

ENUMERATION BASIS: De jure.

GLOSSARY

ADMINISTRATIVE, GEOGRAPHICAL, AND POLITICAL UNITS

Level	English Translation	Vernacular Term
National	Kingdom of Sweden	Konungariket Sverige
Intermediate	County	Län
Local	Commune/community	Kommun
	Market town	Köping
	Municipal community	Municipalsamhälle
	Rural community	Landskommun
	Town	Stad
Population cluster ("locality") concepts:[3]		
	Densely populated area	Tätbebyggelse
	Locality	Tätort
	Sparsely populated area	Glesbebyggelse
Other concepts:		
	Labour Market Board's A-region	Arbetsmarknadsstyrelsens A-region
	Natural farming area	Naturlig jordbruksområde

---

[2] Volume XI contains information on a sample population consisting of all persons born on the 15th of any month, i.e., about 3.3 percent of the total population.

[3] For a discussion of these concepts, see European Population Censuses: The 1960 Series, U.N. Doc.: ST/CES/3 (1964), p. 108.

GLOSSARY

Newspaper circulation area	Tidningsspridningsområde
Province	Landskap

Ecclesiastical hierarchy:

Diocese	Stift
Rural deanery	Kontrakt
Rectoral district	Pastorat
Parish	Församling
District of parish register	Kyrkobokföringsdistrikt

For administrative purposes, Sweden is divided into counties. The city of Stockholm forms one such county.

Communes/communities are the basic units of local government. There are three kinds: market towns, rural communities, and towns, distinguished by different types of administrative organization. This organization depends on characteristics stemming from population density, but population is not the only determining factor. In certain rural communities, the density of population in specific areas is so great that they assume the characteristics of urban areas. Such areas, called municipal communities, remain part of the rural community, but appropriate agencies are created within them to cope with problems not common to the remaining community.

The administration of market towns, in terms of urban functions, stands between that of urban communities and towns. The latter generally include the important urban areas of Sweden, though, in terms of size alone, many towns have smaller populations than some market towns and urban communities. Many, moreover, embrace rural as well as urban territory, since rural areas surrounding towns are sometimes incorporated into them when the areas have proved incapable of efficient self-government.

Densely populated areas within a specific region (e.g., a county) include all localities and parts of localities within that region. Localities, areas defined irrespective of other divisions and existing solely for census purposes, consist of at least 200 registered inhabitants in houses no farther than 200 meters apart. The distance may vary if the density of surrounding rural areas, or of areas within the locality, would otherwise result in artificial boundaries. Uninhabited buildings, with the exception of farming structures, and in some cases summer houses, are considered in delimiting the boundaries of localities. Open areas wider than 200 meters within a locality are disregarded if used for public purposes, as are bodies of water spanned by bridges. Institutions outside localities are considered to be localities in themselves when the total resident personnel, excluding patients, is 200 or more.

<u>Sparsely populated areas</u> are all areas within a specific region which are not defined as densely populated areas.

<u>Labour Market Board's A-regions</u> are areas around certain centers with service establishments of different types serving principally the whole region.

The country is divided into 60 <u>natural farming areas</u>. These are frequently grouped into 18 production areas, which in turn are grouped into 8 major regional production areas. Stockholm (city and county) forms one such natural farming area, and the boundaries of 6 other counties coincide with the boundaries of natural farming areas. In all other cases, the natural farming areas embrace territory of more than one county.

<u>Newspaper circulation areas</u> are areas in which, generally speaking, the population reads the same local newspaper.

<u>Provinces</u> are historical divisions whose boundaries conform closely to those of the counties.

There are 13 <u>dioceses,</u> which are the largest ecclesiastical units in the country. They are composed of <u>rural deaneries,</u> which are made up of <u>rectoral districts,</u> which in turn are made up of <u>parishes.</u> Parish boundaries closely conform to those of the communes/communities. Certain parishes are further divided into <u>districts of parish register.</u>

<u>TABLE OF DEMOGRAPHIC CONCEPTS</u>

The demographic concepts which appear in the following table correspond to the recommendations of the European Conference of Statisticians, unless otherwise noted. They are defined and discussed in the appendixes and in footnotes. Terms in parentheses are alternative forms which appear in the table titles.

<u>Appendix</u>	<u>English Table Term</u>	<u>Vernacular Table Term</u>
ECONOMIC CHARACTERISTICS		
A	Economic activity	Sysselsättning
	Economically active[1]	Förvärvsarbetande befolkning
	Economically inactive	Ej förvärvsarbetande befolkning

---

[1][All footnotes are at the end of the Table.]

GLOSSARY

	Commuters[2]	Pendlare
	Economically active day-population[3]	Förvärvsarbetande dagbefolkning
B	Employment status	Yrkesställning
C	Industrial classification[4]	Näringsgren
	Division(s) (Industry)	Näringsgren (1-siffernivå)
	Major group(s) (Industry)	Näringsgren (2-siffernivå)
	Group(s) (Industry)	Näringsgren (3-siffernivå)
D	Occupational classification[5]	Yrke
	Major group(s) (Occupation)	Yrke (1-siffernivå)
	Minor group(s) (Occupation)	Yrke (2-siffernivå)
	Unit group(s) (Occupation)	Yrke (3-siffernivå)
E	Socio-economic categories	Socio-ekonomiska grupper

HOUSEHOLD AND FAMILY CHARACTERISTICS

F	Household unit	Hushåll
	Private household	Privata hushåll
	Dwelling household[6]	Bostadshushåll
	Other private household[7]	Andra privata hushåll
	Institutional household[8]	Kollektivhushåll
	Head of household[9]	Hushållsforständere

---

[1] The economically active are persons who, during the census week, were gainfully employed for at least one half of normal working hours, members of the family who assisted gainfully employed persons (even if the family members were unpaid), and people who were temporarily not working because of holidays, illness, or unemployment.

[2] Commuters are all persons who, during the week of October 2-8, 1960, worked outside the commune or rural parish of their registered residence. Included were persons who commuted irregularly or temporarily.

[3] The economically active day-population includes all employed persons in a given commune or rural parish, regardless of the location of their place of residence--that is, the resident population less outgoing commuters (utpendlare), plus incoming commuters (inpendlare).

[4] In the table titles, "industry" designates tabulations of divisions, major groups, or groups. Usage varies.

SWEDEN

[5] In the table titles, "occupation" designates tabulations of major groups, groups, or unit groups.

[6] Dwelling households are those units intended as private residences in which persons are registered as living, including persons employed in the household, boarders, and lodgers.

[7] Other private households are those in which persons with unknown place of residence, wards of the parish, and others who do not normally form a household, in the usual sense of the word, live. The category embraces people not living in a residence dwelling.

[8] Institutional households are those in which persons are registered, such as bachelor homes, hospitals, etc., and which are not intended as private residences.

[9] The relationship to the head of the household was determined for all persons living in private dwelling households. In a dwelling household, the person who, because of ownership, rental contract, or other agreement, has disposal of the dwelling unit as a whole, is considered the head of the household. A dwelling household in which none of the inhabitants disposes of the dwelling in its entirety is not considered to have a head of household. In single-person households, which constitute the majority of other private households, the single person is the head of the household. In households whose occupants are listed as wards of the parish or commune, the husband, or only existing parent, or in the absence of any parents, the eldest member of the household, is regarded as the head of the household.

TABLE TITLES BY VOLUME

I. POPULATION IN COMMUNITIES AND PARISHES BY SEX, AGE, AND MARITAL STATUS, ETC.

TABLES

Number	Title	Page
1.	Population by sex and marital status, area, relative number of population in localities and density of population in communities, parishes and districts of parish register.	1
2.	Population by sex and age in communities, parishes and districts of parish register.	62

TABLE TITLES BY VOLUME

II. POPULATION IN LOCALITIES BY SEX, AGE, AND MARITAL STATUS.

TABLES

Number	Title	Page
1.	Population in localities with at least 200 inhabitants, by sex and marital status.	1
2.	Population in localities with at least 200 inhabitants, by sex and age.	34
3.	Population in certain localities with under 200 but at least 150 inhabitants.[1]	102

III. POPULATION IN THE WHOLE COUNTRY AND IN THE COUNTIES, BY SEX, AGE, AND MARITAL STATUS, ETC.

TABLES

Number	Title	Page
	Tables for the Whole Country	
1.	Population by age (five-year groups), sex, and marital status in rural areas, in towns, and in the whole country.	2
2.	Population by age (one-year groups), sex, and marital status in rural areas, in towns, in the whole country, and in densely and sparsely populated areas.	5
3.	Population by age (real age on November 1, five-year groups), sex, and marital status in the whole country.	13
4.	Population by age (real age on November 1, one-year groups), sex, and marital status in rural areas, in towns, in the whole country, and in densely and sparsely populated areas.	14
	Tables for the Counties	
5.	Population by age (five-year groups), sex, and marital status, by county.	22
6.	Relative distribution of population by age (five-year groups), sex, and marital status, by county.	46
7.	Population in densely and sparsely populated areas by age (five-year groups), sex, and marital status, by county.	54

---

[1]Table 3 presents data on settlements which were thought to fit the building-density criteria for a locality, but which subsequently were found to have only 150-199 registered inhabitants.

8. Population by age (one-year groups) and sex, by county. 78

9. Relative distribution of population by age (one-year groups) and sex, by county. 92

Tables for Larger Towns

10. Population by age (five-year groups), sex, and marital status in towns with at least 25,000 inhabitants. 98

11. Population by age (one-year groups), sex, and marital status in towns with at least 75,000 inhabitants. 109

Tables for Size Groups of Localities, etc.

12. Population by age (five-year groups), sex, and marital status in densely and sparsely populated areas and in localities, by size of locality. 121

13. Relative distribution of population by age (five-year groups), sex, and marital status in densely and sparsely populated areas and in localities, by size of locality. 125

14. Number of localities and population of localities, by size groups, by county. 129

Table for Married Women, by Year of Marriage

15. Married women by age and year of marriage in rural areas, in towns, in the whole country and in densely populated areas. 130

IV. POPULATION IN DIFFERENT DIVISIONS,[1] ALIENS, ETC.

TABLES

Number	Title	Page

Regional Tables

1. Counties: Population, area, and density of population

---

[1] Volume IV includes several administrative and political divisions not defined in the glossary. The county council district (landstingsområde) includes the same territory as the county outside the largest cities. Each county usually contains only one county agricultural committee district (hushållningssällskapsområde), though a few contain more than one. The country is also divided for judicial purposes into circuits (domsagor), some of which are further divided into assizes (tingslag). For law enforcement purposes, the country is divided into constabulary districts (landsfiskaldistrikt); for assessment and tax purposes, into national registration districts (fögderier).

	in rural communes, market towns, towns, and in the whole country.	1
2.	Counties: Population by sex and marital status in the whole county and in densely populated areas.	2
3.	Counties: Population by sex and age in the whole county and in densely populated areas.	4
4.	Provinces: Population, area, and density of population.	8
5.	County council districts: Population, area, and density of population.	9
6.	County agricultural committee districts: Population, area, and density of population.	10
7.	Circuits and assize division: Population, area, and density of population.	11
8.	National registration districts: Population, area, and density of population.	20
9.	Constabulary districts: Population, area, and density of population.	29
10.	Municipal communities: Population, area, and density of population.	40
11.	Dioceses, rural deaneries, and rectorial districts: Population and area.	42
12.	Parishes that are divided over two or more communities or districts of parish register: Population by sex and marital status, area, relative number of population in localities and density of population.	67
13.	Parishes that are divided over two or more communes or districts of parish register: Population by sex and age.	70
14.	Parishes by relative number of population in localities: Number of parishes and population by sex and age and by county.	74
15.	Natural farming areas: Population by sex and marital status, area and density of population.	82
16.	Natural farming areas: Population by sex and age.	84
17.	Labour Market Board's A-regions: Population by sex and marital status, area and density of population.	88
18.	Labour Market Board's A-regions: Population by sex and age.	90
19.	Newspaper circulation areas: Population by sex and marital status, area and density of population.	94

SWEDEN

20.	Newspaper circulation areas: Population by sex and age.	96
21.	Islands: Population, area, and density of population.	100
22.	Lakes: Area and altitude of the water surface.	104

Persons Born Abroad and Aliens

23.	Swedish citizens and aliens by country of birth and by sex, etc.	106
24.	Persons born abroad and aliens by country of birth or citizenship respectively, and by sex.	107
25.	Persons born abroad by sex, age, and marital status, and by country of birth.	108
26.	Aliens by sex, age, and marital status, and by citizenship.	110
27.	Aliens in different counties and larger towns by citizenship.	112

In the "Residence Unknown" Register: Remaining Persons[2]

28.	In the "residence unknown" register, remaining persons by year of birth and marital status (at the transfer), and by sex, etc.	113
29.	In the "residence unknown" register, remaining persons by sex and year of transfer, and by age at the transfer.	114
30.	In the "residence unknown" register, remaining persons by citizenship and by year of transfer.	115
31.	In the "residence unknown" register, remaining persons by counties (at the transfer) and by sex and year of transfer.	116

---

[2] The "residence unknown" category includes mainly persons without known residence who have been absent from their home commune/ community for at least two years. If no later information is received (e.g., that a person has died or settled abroad), he is not deleted from the register until he reaches the age of 100. Most such persons, however, must be supposed to have died, even if no confirmation of this has been received. Others in this category are aliens who registered in Sweden but who left the country without reporting to the authorities.

VI. ECONOMICALLY ACTIVE POPULATION BY INDUSTRY AND STATUS, ETC., IN COMMUNES, PARISHES, AND LOCALITIES.

TABLES

Number	Title	Page
1.	Economically active population by industry and economically inactive population in communes and parishes.	6
2.	Economically active population by industry, employment status, and age, in communes.	140
3.	Economically active population by industry and economically inactive population in localities with at least 200 inhabitants.	298
4.	Economically inactive population by age and members of families, economically inactive, supported by head of the family, by industry, in communes.	370

VII. ECONOMICALLY ACTIVE POPULATION BY OCCUPATION AND HOUSEHOLDS BY SIZE, ETC., WITHIN COMMUNES AND LARGER LOCALITIES.

TABLES

Number	Title	Page
1.	Economically active population, by occupation and sex, in communes.	6
2.	Economically active population, by occupation and sex, in localities with at least 1,000 inhabitants.	102
3.	Population, by type of household and position in household, in communes.	164
4.	Private households, by number of members and possession of car, and private households, with married head of household living with his wife, by number of children under 16 years of age, in communes.	187
5.	Private households, by number of members and possession of car, and private households, with married head of household living with his wife, by number of children under 16 years of age, in localities with at least 2,000 inhabitants.	210

SWEDEN

VIII. ECONOMICALLY ACTIVE DAY-POPULATION BY INDUSTRY AND OUT-GOING AND INGOING COMMUTING IN COMMUNES AND RURAL PARISHES.

TABLES

Number	Title	Page
1.	Out- and in-going commuting and size of economically active day-population, in communes and rural parishes.	6
2.	Economically active day-population by industry, in communes.	70
3.	Economically active outgoing commuters, commuting over commune boundary, by industry and age, in communes with at least 100 outgoing commuters.	116
4.	Economically active ingoing commuters, commuting over commune boundary, by industry and age, in communes with at least 100 ingoing commuters.	135
5.	Economically active ingoing commuters by occupation, in towns with at least 500 ingoing commuters.	148
6.	Economically active outgoing commuters by place of work (rural parish or town), in communes with at least 100 outgoing commuters.	158
7.	Economically active ingoing commuters by place of residence (rural parish or town), in communes with at least 100 ingoing commuters.	188
8.	Private households by outgoing commuters, commuting over commune boundary, in communes with at least 100 outgoing commuters.	215

IX. INDUSTRY, OCCUPATION, COMMUTING, HOUSEHOLDS, AND EDUCATION IN THE WHOLE COUNTRY, BY COUNTY, ETC.

TABLES

Number	Title	Page
	Economically Active and Economically Inactive Population	
1.	Population by type of economic activity, sex, and age, in the whole country.	6
2.	Economically active and economically inactive population by sex, age, and marital status, in the whole country.	8

Industry, Employment Status, and Kind of Employment[1]

3. Economically active population and members of families of economically active persons by industry and employment status and by sex, in the whole country and in densely populated areas. — 10

4. Economically active population by industry and employment status and by sex, age, and marital status, in the whole country. — 22

5. Economically active population by industry, sex, age, and marital status and economically inactive population, in the whole country, in sparsely and densely populated areas, and in rural areas and towns. — 38

6. Economically active population by industry and sex in sparsely and densely populated areas, in rural areas, in towns, and in the whole country. — 48

7. Economically active population by industry, employment status, and kind of employment, and by sex, in the whole country. — 50

8. Married persons living together by type of economic activity and industry of the husband and the wife, respectively, in the whole country. — 52

9. Family workers by industry and employment status and by sex, age, and marital status, in the whole country. — 54

10. Members of families, economically inactive, supported by head of the family, by industry and employment status, in the whole country. — 55

11. Economically active population by industry and economically inactive population, in sparsely and densely populated areas, by county. — 56

12. Economically active population by employment status, kind of employment, and sex, economically inactive members of families and remaining economically inactive population, by county, with the distinction of densely populated areas. — 60

13. Economically active population by industry, employment status, and sex, and economically inactive members of families by industry, by county, with the distinction of densely populated areas. — 64

---

[1] Kind of employment classifies employees into governmental, communal, and private services.

14.	Economically active population by industry, sex, and age, and economically inactive population by sex and age, by county.	70
15.	Economically active population by industry and economically inactive population, in A-regions.	82
16.	Economically active population by industry and economically inactive population, in newspaper circulation areas.	86
17.	Economically active population by industry and economically inactive population, in natural farming areas.	90
18.	Economically active population by industry, sex, and age, and economically inactive population by sex and age, in towns with at least 50,000 inhabitants.	94
19.	Total population and economically active population by industry, in the principal localities[2] of the towns and in suburbs.	102
20.	Population in different types of parishes[2] by relative size of population in densely populated areas, and by the share of the economically active population in agriculture, forestry, etc., and in manufacturing, etc., in the whole country.	108
21.	Population in different types of parishes[2] by the share of the economically active population in agriculture, forestry, etc., and in manufacturing, etc., by county.	110
22.	Population in different types of localities[2] by the share of the economically active population in agriculture, forestry, etc., and in manufacturing, etc., by county.	115

Occupation and Socio-economic Groups

23.	Economically active population by occupation, sex, and age, and employees by kind of employment,[1] in the whole country.	118
24.	Economically active population by occupation, employment status, and sex, in the whole country.	130
25.	Economically active population by occupation, employment status, and sex, in the whole country and in densely populated areas.	136

---

[2] Classification by type of parish or locality refers to the predominant industrial character of the area.

26.	Economically active population by occupation and industry, in the whole country.	138
27.	Married persons living together by type of economic activity and occupation of the husband and the wife respectively, in the whole country.	150
28.	Economically active population by occupation and sex, by county with the distinction of densely populated areas.	152
29.	Economically active population by occupation, in A-regions.	160
30.	Economically active population and economically inactive members of families by socio-economic groups and by sex, age, and marital status, in the whole country and in densely populated areas.	168
31.	Economically active population and economically inactive members of families by socio-economic groups, by county, with the distinction of densely populated areas.	170

Commuting

32.	Economically active population by place of residence and place of work, and by industry and sex, in the whole country.	172
33.	Economically active population by place of work, by county, with distinction of rural areas.	174
34.	Economically active night-population and day-population by industry, in Greater Stockholm and Greater Gothenburg.	176
35.	Economically active commuters, commuting over commune boundary, by industry, sex, and age, in the whole country.	178
36.	Economically active commuters, commuting over commune boundary, by occupation and age, in the whole country.	180

Households

37.	Population by sex, age, and marital status, and by type of household and relationship to head of household, in the whole country and in densely populated areas.	182
38.	Population by type of household and relationship to head of household, by county with the distinction of densely populated areas.	184
39.	Private households by sex, age, and marital status of head of household, and by number of members, etc., in the whole country.	186

SWEDEN

40. Private households by sex, age, and marital status of head of household, and by number of members and number of children under 16 years of age, in the whole country and in densely populated areas. 188

41. Private households, with married head of household, by duration of marriage, by number of members and by economic activity and age of wife, in the whole country and in densely populated areas. 192

42. Private households by sex, industry, and employment status of head of household, and by number of members and number of children under 16 years of age, in the whole country. 194

43. Private households by age of head of household, and by number of members, etc., by county. 198

44. Private households by age of head of household, and by number of members, etc., in towns with at least 30,000 inhabitants. 202

45. Private households by number of members, and by number of outgoing commuters, commuting over commune boundary, in the whole country and in rural areas and towns. 204

46. Private households by type of household, by economic activity of head of household, by number of adults and number of children under 16 years of age, in the whole country and in sparsely and densely populated areas.[3] 206

47. Private households by type of household and number of members, and by relationship of the members to head of household, in the whole country.[3] 207

48. Private households by occupation of head of household and by number of members and number of children under 16 years of age, in the whole country.[3] 208

49. Private households with married head of household living with his wife, by economic activity and occupation of the wife, by number of their children under 16 years of age, and by age of the youngest child, in the whole country.[3] 210

50. Private households with economically active head of household, by socio-economic group of head of household, and by number of members and number of children under 16 years of age, in the whole country and in sparsely and densely populated areas.[3] 211

---

[3] Based on a ten percent sample.

Education[4]

51. Private households, in which at least one member has an academic degree, by age and industry of head of household, and by number of members of household, by number of children, etc., in the whole country. 214

52. Economically active and economically inactive persons with certain kinds of education by sex and age, in the whole country. 216

53. Economically active persons with certain kinds of education by industry, kind of employment and sex, in the whole country. 224

54. Economically active persons with certain kinds of education by occupation and sex, in the whole country. 226

55. Economically active and economically inactive persons with certain kinds of education by sex, by county. 230

56. Persons with certain kinds of education by sex and age, in towns with at least 50,000 inhabitants. 232

X. MAIN OCCUPATION, ARABLE AND FOREST ACREAGE, PRIVATE CAR OWNERS, AND NATIONALITY.

TABLES

Number	Title	Page

Main Occupation

1. Economically active population during the year Oct. 2, 1959-Oct. 1, 1960 compared with the week Oct. 2-8, 1960 in communes. 4

2. Economically active population by occupation during the year Oct. 2, 1959-Oct. 1, 1960 and by sex, by county. 16

3. Economically active population by occupation partly during the week Oct. 2-8, 1960 and partly during the year Oct. 2, 1959-Oct. 1, 1960 and by sex, by county. 20

---

[4] Educational level refers to the kind of examination passed. Persons who have passed more than one examination have generally been classified under the highest examination, or, when the examinations are of the same level, under the examination most closely associated with the individual's occupation.

SWEDEN

4. Economically active population by occupation partly during the week Oct. 2-8, 1960 and partly during the year Oct. 2, 1959-Oct. 1, 1960 and by sex, in towns with at least 50,000 inhabitants. ... 25

5. Economically active population by occupation and economically inactive population partly during the week Oct. 2-8, 1960, partly during the year Oct. 2, 1959-Oct. 1, 1960, in the whole country. ... 28

6. Economically active population by occupation during the year Oct. 2, 1959-Oct. 1, 1960 and by sex and age, in the whole country. ... 34

7. Persons with different occupation or activity during the year Oct. 2, 1959-Oct. 1, 1960 than during the week Oct. 2-8, 1960, by sex and age, in the whole country, sparsely and densely populated areas. ... 36

8. Persons with different occupation during the year Oct. 2, 1959-Oct. 1, 1960 than during the week Oct. 2-8, 1960 by industry and by sex and age, in the whole country. ... 37

9. Persons with different occupation or activity during the year Oct. 2, 1959-Oct. 1, 1960 than during the week Oct. 2-8, 1960 by occupation or activity and by sex and age, in the whole country. ... 38

Population Living on Farm Real Estates

10. Economically active population by industry and economically inactive population, living on farm real estates, in communes. ... 39

11. Economically active population by industry and sex and economically inactive population by sex, living on farm real estates, by county. ... 62

12. Economically active population by industry and economically inactive population living on farm real estates, in natural farming areas. ... 64

13. Economically active population by industry and economically inactive population living on farm real estates, in A-regions. ... 68

14. Population, living on farm real estates, by type of activity, sex, age, and marital status, in the whole country. ... 72

15. Economically active population, living on farm real estates, by industry, sex and age, in the whole country. ... 73

16. Economically active population, living on farm real estates, by occupation during the year Oct. 2, 1959-

	Oct. 1, 1960 and by sex and marital status, in the whole country.	74
	Farmers, Tenants, etc.	
17.	Economically active and economically inactive farmers according to the National Farmers' Union's farm register, by arable acreage and age, in natural farming areas.	75
18.	Members of households of farmers according to the National Farmers' Union's farm register, by industry, arable acreage, and sex, in natural farming areas.	83
19.	Economically active farmers according to the National Farmers' Union's farm register, by industry, arable and forest acreage, and age, in the whole country.	91
20.	Farmers according to the National Farmers' Union's farm register, by occupation during the year Oct. 2, 1959-Oct. 1, 1960 and by arable and forest acreage and age, in the whole country.	93
21.	Members of households of farmers according to the National Farmers' Union's farm register, by occupation during the year Oct. 2, 1959-Oct. 1, 1960 and by arable and forest acreage, sex, and age, in the whole country.	95
	Private Car Owners	
22.	Private car owners by sex and in relation to the population 18 years of age and over, in communes.	97
23.	Economically active private car owners by industry and economically inactive private car owners, by county and in towns with at least 50,000 inhabitants.	109
24.	Private car owners by sex and marital status and in relation to the population 18 years of age and over, in the whole country, sparsely and densely populated areas.	110
25.	Economically active private car owners by industry, status, sex and age, and economically inactive private car owners by sex and age, in the whole country.	111
26.	Economically active private car owners by occupation and sex, in the whole country, and in densely populated areas.	112
	Persons Born Abroad and Aliens	
27.	Economically active persons born abroad by industry, status, and sex, and economically inactive persons born abroad by sex, by county, and in towns with at least 50,000 inhabitants.	118

SWEDEN

28.	Economically active persons born abroad by industry, sex, and age, in the whole country.	120
29.	Economically active persons born abroad by country of birth, industry, and sex, in the whole country.	122
30.	Economically active persons born abroad by occupation, sex, and age, in the whole country.	124
31.	Economically active aliens by industry, status, and sex, and economically inactive aliens by sex, by county, and in towns with at least 50,000 inhabitants.	130
32.	Economically active aliens by industry, sex, and age, in the whole country.	132
33.	Economically active aliens by citizenship, industry, and sex, in the whole country.	134

XI. SAMPLE SURVEYS: FAMILIES, INCOME, INTERNAL MIGRATION, AND CHANGE OF INDUSTRY.[1]

TABLES

Number	Title	Page
	Families[2]	
1.	Families by type of family and by number of children, in the whole country and in densely populated areas.	3
2.	Families by sex, age, and marital status of head of family, and by number of children under 16 years of age and under school age (0-6 years of age), in the whole country and in densely populated areas.	4
3.	Families by sex, age, and marital status of head of family, and by age of the youngest child, in the whole country.	6

---

[1] This volume presents data from the population register on that sample of the population born on the 15th of every month--approximately 3.3 percent of the total population.

[2] A family is a group of at least two persons registered within the same household, living together with or without children under 16 years of age who live at home; or an unmarried, previously married, or married person not living with his or her spouse, together with children under 16 years of age living at home. The husband in an existing marriage, or the parent in families where there is no husband present, is regarded as the head of the family.

TABLE TITLES BY VOLUME

4. Families by sex, age, marital status, and economic activity of head of family, and by number of children under 16 years of age and under school age (0-6 years of age), in the whole country and in densely populated areas.   8

5. Families with economically active head of family, by sex, age, marital status, occupation, and status of head of family, in the whole country.   10

6. Families with economically active head of family, by sex, marital status, occupation, and status of head of family, and by number of children under 16 years of age and under school age (0-6 years of age), in the whole country.   12

7. Marriages by year of marriage, number of children under 16 years of age and under school age (0-6 years of age), in the whole country and in densely populated areas.   14

8. Marriages by age of husband and of wife, in the whole country.   16

9. Marriages by economic activity of husband and of wife, and by number of children under 16 years of age and under school age (0-6 years of age), in the whole country and in densely populated areas.   17

10. Marriages with children, by age of wife, economic activity of husband and of wife, and by age of the youngest child, in the whole country.   18

11. Marriages by year of marriage, occupation, and status of husband and of wife, in the whole country.   20

12. Marriages performed in 1951 or later by year of marriage, number of marriage, age of wife when married and by number of children and their years of birth, in the whole country.   28

13. Married women living with their husbands, by age, year of marriage, and by number of children under 16 years of age, in the whole country.   34

14. Economically active married women living with their husbands, by their own age, by age of the youngest child, and by number of children under 16 years of age, in the whole country.   35

15. Economically active women (excl. married women living with their husbands) by their own age, by age of the youngest child, and by number of children under 16 years of age, in the whole country.   36

16. Economically active women by occupation and by number of children under 16 years of age and under school age (0-6 years of age), in the whole country.   37

SWEDEN

17. Women born 1921-1930 by age and marital status, number of children born, and their years of birth and legitimacy, in the whole country. 38
18. Children under 16 years of age belonging to families, by age, and by sex and marital status of head of family, and by economic activity of parents, in the whole country. 40

Income

19. Economically active and economically inactive adult population, by sex, age, and income, in the whole country. 42
20. Economically active and economically inactive adult population, by sex, marital status, and income, in sparsely and densely populated areas and in the whole country. 46
21. Economically active adult population, by sex, industry, status, and income, in the whole country. 50
22. Number of economically active income earners and size of median and quartile income, by sex, age, industry, and status, in the whole country. 54
23. Number of economically active income earners and size of median and quartile income, by sex, industry, and status, in sparsely and densely populated areas. 56
24. Number of economically active income earners and size of median and quartile income, by sex and industry, in the whole country. 57
25. Number of economically active income earners, by industry and status, partly by classification of the census of the population and partly by classification of the annual income statistics, in the whole country. 58
26. Number of economically active income earners and size of median and quartile income, by sex, age, and occupation, in the whole country. 60
27. Economically active and economically inactive married persons living together, by income of the husband and the wife, respectively, in the whole country. 74
28. Married persons living together, by industry and income of the husband and the wife, respectively, in the whole country. 76
29. Number of economically active income earners and size of median and quartile income, by type of family and by number of children under 16 years of age, in the whole country and in densely populated areas. 80

30.	Number of mothers (with children under 16 years of age) with income and size of median and quartile income, by marital status of the mother and by age of the youngest child, in the whole country and in densely populated areas.	82
31.	Number of economically active income earners and size of median and quartile income, by certain kinds of education and by age, in the whole country.	83
32.	Number of economically active income earners and size of median and quartile income, by sex, age, and socio-economic group, in the whole country.	84
33.	Number of economically active foreign income earners and size of median and quartile income, by sex, industry, and status, in the whole country.	86
34.	Number of economically active income earners born abroad and size of median and quartile income, by sex, industry, and status, in the whole country.	87

Migration

35.	Population, by sex and county of residence, in 1950 and in 1960.	88
36.	Migration between communes, by county, 1951-1960.	92
37.	Migration from densely and sparsely populated areas, by county, 1951-1960.	93
38.	Migration from densely and sparsely populated areas, by sex and age of the migrants, 1951-1960.	94
39.	Migrants to densely populated areas of different sizes and to sparsely populated areas, by type of parish, 1951-1960.	95
40.	Migrants to densely populated areas of different sizes and to sparsely populated areas, by type of parish and by sex and age of the migrants, 1951-1960.	96
41.	Migrants by county, 1951-1960, by sex and age.	98
42.	Population remaining in the commune and migrants to the commune, 1951-1960, by sex, age, and marital status, in densely and sparsely populated areas.	100
43.	Population remaining in the commune and migrants to the commune, 1951-1960, by type of activity, sex, and age.	104
44.	Population remaining in the commune and migrants to the commune, 1951-1960, by sex and status.	105
45.	Population remaining in the commune and migrants to the commune, 1951-1960, by industry (agriculture,	

etc., and non-agricultural industries), sex, and status. 106

Change of Industry

46. Population by industry (agriculture, etc., and non-agricultural industries), 1950 and 1960, and by sex and age. 107

S W I T Z E R L A N D

MAIN ENTRY AND VOLUME NOTES:

Switzerland. Statistisches Amt.
   Eidgenössische Volkszählung. Recensement fédéral de la population. 1960. Bern, 1961-1966.
   32 v. 30cm.

Band 1.	Wohnbevölkerung der Gemeinden, 1850-1960. 1er volume. Population résidente des communes, 1850-1960. 1961. 87 p.
Band 2.	Kanton Aargau. 1964. 157 p.
Band 3.	Kanton Appenzell Ausser-Rhoden. 1963. 107 p.
Band 4.	Kanton Appenzell Inner-Rhoden. 1963. 103 p.
Band 5.	Kanton Basel-Landschaft. 1963. 133 p.
Band 6.	Kanton Basel-Stadt. 1963. 175 p.
Band 7.	Kanton Bern--Canton de Berne. 1964. 351 p.
8me Volume.	Canton de Fribourg--Kanton Freiburg. 1964. 171 p.
9me Volume.	Canton de Genève. 1963. 191 p.
Band 10.	Kanton Glarus. 1963. 109 p.
Band 11.	Kanton Graubünden. 1964. 147 p.
Band 12.	Kanton Luzern. 1964. 191 p.
13me Volume.	Canton de Neuchâtel. 1964. 135 p.
Band 14.	Kanton St. Gallen. 1964. 194 p.
Band 15.	Kanton Schaffhausen. 1963. 115 p.
Band 16.	Kanton Schwyz. 1963. 111 p.
Band 17.	Kanton Solothurn. 1963. 143 p.
Band 18.	Kanton Thurgau. 1964. 151 p.
19° Volume.	Cantone Ticino. 1963. 155 p.
Band 20.	Kanton Unterwalden nid dem Wald. [Nidwalden] 1963. 105 p.
Band 21.	Kanton Unterwalden ob dem Wald. [Obwalden] 1963. 105 p.
Band 22.	Kanton Uri. 1963. 107 p.

SWITZERLAND

23me Volume.   Canton du Valais--Kanton Wallis. 1964. 153 p.
24me Volume.   Canton de Vaud. 1964. 254 p.
Band 25.   Kanton Zug. 1963. 107 p.
Band 26.   Kanton Zürich. 1963. 293 p.
Band 27.   Schweiz. Teil I. Geschlecht, Heimat, Geburtsort, Konfession, Muttersprache, Zivilstand, Alter, Schulbesuch. 27me volume. Suisse. 1ere partie. Sexe, origine, lieu de naissance, religion, langue maternelle, état civil, âge, formation scholaire. 1964. 187 p.
Band 28.   Schweiz. Teil II. Erwerb und Beruf. 28me volume. Suisse. IIme partie. Branches économiques et professions. 1965. 377 p.
Band 29.   Schweiz. Teil III. Wohnungen. 29me volume. Suisse. IIIme partie. Logements. 1964. 319 p. [Excluded.]
Band 30.   Schweiz. Teil IV. Wohn- und Arbeitsort der Berufstätigen (Pendelwanderung). 30me volume. Suisse. IVme partie. Domicile et lieu de travail des personnes exerçant une profession (migrations alternantes). 1965. 97 p.
Band 31.   Schweiz. Teil V. Heimatgemeinden der Schweizerbürger. 31me volume. Suisse. Vme partie. Communes d'origine des ressortissants suisses. 1965. 86 p.
Band 32.   Schweiz. Teil VI. Haushaltungen. 32me volume. Suisse. VIme partie. Ménages. 1966. 129 p.

ENGLISH TRANSLATION OF CENSUS AND VOLUME TITLES:

General population census of Switzerland, December 1, 1960.

1. Resident population of the communes, 1850-1960.
2. Kanton Aargau.
3. Kanton Appenzell Ausser-Rhoden.
4. Kanton Appenzell Inner-Rhoden.
5. Kanton Basel-Landschaft.
6. Kanton Basel-Stadt.
7. Kanton Bern--Canton de Berne.
8. Canton de Fribourg--Kanton Freiburg.
9. Canton de Genève.
10. Kanton Glarus.

GLOSSARY

11. Kanton Graubünden.
12. Kanton Luzern.
13. Canton de Neuchâtel.
14. Kanton St. Gallen.
15. Kanton Schaffhausen.
16. Kanton Schwyz.
17. Kanton Solothurn.
18. Kanton Thurgau.
19. Cantone Ticino.
20. Kanton Unterwalden nid dem Wald.
21. Kanton Unterwalden ob dem Wald.
22. Kanton Uri.
23. Canton du Valais--Kanton Wallis.
24. Canton de Vaud.
25. Kanton Zug.
26. Kanton Zürich.
27. Sex, origin, birthplace, religion, mother tongue, marital status, age, educational level.
28. Industry and occupations.
29. Housing.
30. Residence and workplace of economically active population (commuters).
31. Communes of origin of Swiss citizens.
32. Households.

ENUMERATION DATE: December 1, 1960.

ENUMERATION BASIS: De jure.

GLOSSARY

ADMINISTRATIVE, GEOGRAPHICAL, AND POLITICAL UNITS

Level	English Translation	Vernacular Term
National	Switzerland	Schweiz, Suisse, Svizzera
Intermediate	Canton	Kanton, canton, cantone

SWITZERLAND

Local	Commune	Gemeinde, commune, comune
Population cluster ("locality") concepts:*		
	Population aggregate	Agglomeration, agglomération, agglomerazione
	Town or city	Stadt, ville, città

The Republic of Switzerland comprises 22 <u>cantons</u>. The cantons are sovereign in all fields of government except where the Constitution expressly provides for the jurisdictional competence of the Republic.

Cantons are composed of <u>communes</u>, which are autonomous units with the right of representative government. The extent of autonomy allotted to communes varies from canton to canton.

Communes with populations of more than 10,000 persons are called <u>towns</u> or <u>cities</u> for statistical purposes.

<u>Population aggregates</u> are densely populated areas containing a central town and suburbs.

TABLE OF DEMOGRAPHIC CONCEPTS

The demographic concepts which appear in the following table correspond to the recommendations of the European Conference of Statisticians, unless otherwise noted. They are defined and discussed in the appendixes and in footnotes. Terms in parentheses are alternative forms which appear in the table titles. Vernacular terms are presented, respectively, in German, French, and Italian.

<u>Appendix</u>	<u>English Table Term</u>	<u>Vernacular Table Terms</u>
CULTURAL, EDUCATIONAL, AND PERSONAL CHARACTERISTICS		
...	Educational level[1]	Schulstufe Degré scolaire (degré d'enseignement) [No Italian equivalent]
	Mother tongue[2]	Muttersprache Langue maternelle Lingua materna

---

*For a discussion of these concepts, see <u>European Population Censuses: The 1960 Series</u>, U.N. Doc.: ST/CES/3 (1964), p. 108.

[1] [All footnotes are at the end of the Table.]

GLOSSARY

	Origin[3]	Heimat Origine Attinenza
	Religion[4]	Konfession Religion Religione

ECONOMIC CHARACTERISTICS

A	Economic activity	Berufstätigkeit Activité professionnelle [No Italian equivalent]
	Economically active	Aktive Bevölkerung (Berufstätige) Population active (personnes exerçant une profession) Popolazione attive (persone esercitanti una professione)
	Economically inactive	Nicht aktive Bevölkerung Population non active Popolazione inattive
B	Employment status	Beruflicher Stellung Situation professionnel Posizione nella professione
C	Industrial classification Division(s)	Branche économique[5] Erwerbsklassen Classes économiques Classi economiche
	Major group(s)	Erwerbsgruppen Groupes économiques Gruppi economici
	Group(s)	Erwerbsarten Branches économiques Rami economici
D	Occupational classification[6]	Persönlicher Beruf Profession individuelle Professione individuale
E	Socio-economic category	Socio-ökonomische Gruppe Groupe socio-économique Gruppo socio-economica
F	Household(s)[7]	Haushaltungen Ménages Economie domestiche
	Private household(s)[8]	Privathaushaltungen Ménages privés

...

SWITZERLAND

Family household(s)[9]	Familienhaushaltungen Ménages familiaux
Non-family household(s)	Nichtfamilienhaushaltungen Ménages non familiaux
Institutional household(s)[10]	Kollektivhaushaltungen Ménages collectifs Economie domestiche collettive

---

[1] Educational level indicates the level of the last school attended as a full-time student, currently or in the past.

[2] Mother tongue is that language in which one thinks and can best express oneself. For children who cannot speak yet, the mother's language is indicated. Categories of language are: German, French, Romanche, and other.

[3] Origin of a person is his nationality at the time of the census; of persons of Swiss nationality, the commune and canton of citizenship.

[4] Religion comprises the following categories: Protestant, Roman Catholic, Christian Catholic, Jew, other religion, and without religion.

[5] Volume 28, Switzerland, Part II, "Industry and Occupations," is in French. The German and Italian equivalents for the levels of industrial classification have been included here because they are employed in table titles in other volumes of the census.

[6] Throughout the table titles, "occupation" designates tabulations of one, two, or all three occupational groups (major groups, minor groups, unit groups).

[7] The classification by household employs a national definition based on the household dwelling concept recommended by the European Conference. Households comprise people living in the same dwelling at the time of the census. There are two types of households: private and institutional. Family members living with other families in other homes or in schools were enumerated in the location in which they were living.

[8] Private households are composed of the members of one family living together and of other people who may live with the family. People who live alone, or people living together who are not related by family, constitute a household when they maintain either a dwelling of their own or a room set apart from the dwelling of their landlord. For the census, each private household is considered a dwelling.

[9] The classification of private households as "family" and "non-family" employs a national definition, given below, based on the concept of related members of a household recommended by the European Conference.

Family households
  Family nucleus of the head of the household
  Family of the head of the household plus other relatives
  Family of the head of the household plus non-family members
  Family of the head of the household plus other relatives and non-family members

Non-family households
  Head of the household only
  Head of the household and related persons
  Head of the household and unrelated persons
  Head of the household and related and unrelated persons

In family households the kinds of family nuclei are: head of the household, spouse, and children; head of the household and children; and head of the household and father and/or mother. Other family nuclei living in a household (e.g., a brother of the head of the household and his wife and children, or family domestics) are not counted as independent household units, but rather as relatives or people outside the family of the head of the household who live in the household. Non-family households are private households of one person or of several persons, where none is a spouse, child, mother, or father of the head of the household. The members of private households are classified by status in the household (Stellung in der Haushaltung, situation dans le ménage). In each household one person is considered the head of the household. The status of other members of the household is determined by their relation to the head of the household, kinship having priority (e.g., as where a nephew is also an apprentice).

[10] Institutional households are composed of people living in institutions, hotels, boarding houses, hospitals, barracks, etc. This classification also includes those persons not classified in private households. There is no head of the household in institutional households.

TABLE TITLES BY VOLUME

VOLUME 1. RESIDENT POPULATION OF THE COMMUNES, 1850-1960.

TABLES

Number	Title	Page
1.	Resident population of cantons, 1850-1960.	11
2.	Resident population of communes, 1850-1960.	13

SWITZERLAND

3.	Zürich.	15
4.	Bern.	19
5.	Luzern.	30
6.	Uri.	32
7.	Schwyz.	32
8.	Obwalden.	33
9.	Nidwalden.	33
10.	Glarus.	33
11.	Zug.	34
12.	Fribourg.	34
13.	Solothurn.	40
14.	Basel-Stadt.	42
15.	Basel-Landschaft.	42
16.	Schaffhausen.	44
17.	Appenzell Ausser-Rhoden.	45
18.	Appenzell Inner-Rhoden.	45
19.	St. Gallen.	46
20.	Graubünden.	48
21.	Aargau.	54
22.	Thurgau.	59
23.	Ticino.	64
24.	Vaud.	69
25.	Valais.	77
26.	Neuchâtel.	81
27.	Genève.	83

VOLUMES 2-26. KANTON AARGAU. . .KANTON ZÜRICH.

Structure of the Volumes

Volumes 2-26 are composed of two sets of data per volume: "Highlights of Population Censuses from 1850-1960" and "Results of the Census of December 1, 1960."

Identical table numbers, titles, and pages are found throughout Volumes 2-26 for "Highlights of Population Censuses from 1850-1960"; hence the example provided below applies to all of the volumes.

TABLE TITLES BY VOLUME

For "Results of the Census of December 1, 1960," however, only the table numbers and titles are uniform, and page locations vary. Hence it is necessary to consult the Tables list and the Page Location Matrix on pp. 394-397 in order to locate desired data.

TABLES

Number	Title	Page
	Highlights of Population Censuses from 1850-1960	
1.	Resident population of Switzerland and of the cantons since 1850.	19
2.	Resident population of the communes since 1850.	42[1]
3.	Inhabited buildings and households since 1860.	19
4.	Residences and inhabitants by residence 1920, 1950, and 1960.	20
5.	Components of change in the Swiss population for intercensal periods since 1870.	20
6.	Resident population by sex and origin since 1850.	20
7.	Resident population by sex and marital status since 1860.	21
7a.	Swiss population by sex and marital status since 1900.	21
7b.	Foreign population by sex and marital status since 1900.	21
8.	Population of marriageable age by sex and marital status since 1860.	22
9.	Resident population by age since 1880.	22
9a.	Resident population by age and sex since 1880.	23
	Resident population by age and marital status since 1880:	
10.	Entire population.	24
10a.	Men.	26
10b.	Women.	27
11.	Resident population by origin since 1860.	28
12.	Foreigners by country of origin since 1860.	29
12a.	Foreigners by country of origin and sex since 1880.	29
13.	Resident population by canton of birth since 1888.	30
14.	Resident population by religion since 1860.	30
14a.	Resident population by religion, origin, and sex since 1910.	31

---

[1] Because of the length of the tabulations in Table 2, it is always placed at the end of the series of retrospective tables to allow space for inclusion of all the data.

SWITZERLAND

15.	Religion of spouses living together since 1880.	31
16.	Resident population by mother tongue and origin since 1880.	32

Resident population by economic dependence since 1888:

17.	Entire population.	33
17a.	Men.	34
17b.	Women.	35

Economically active population and economically inactive members of their families by industrial division since 1888:

18.	Absolute numbers.	36
18a.	Numbers per thousand.	37
19.	Economically active men by industrial division since 1888.	38
19a.	Economically active women by industrial division since 1888.	39
20.	Economically active population by industrial division, origin, and sex since 1910.	40
21.	Agricultural population since 1888.	41

Results of the Census of December 1, 1960

1. General table.
2. Households; resident population by commune, sex, marital status, birthplace, origin, religion, and mother tongue.
3. Resident population by commune, age group, and sex; economically active population by commune, industrial division, and sex.
4. Inhabited buildings and inhabitants by commune.
5. Resident population by canton of origin, country of origin, and sex.
6. Resident population by canton of birth, country of birth, origin, and sex.
7. Resident population by year of birth, sex, origin, and marital status.
8. Resident population by five-year age group, sex, origin, and marital status; Canton of _____ [plus principal communes and/or aggregates within the named canton].
9. Economically active persons and their relatives by industrial group.

TABLE TITLES BY VOLUME

9a. Staff, their relatives, and inmates of institutions by origin and kind of institution.

10. Economically active population by major industrial group and employment status.

11. Family members of economically active persons by major industrial group and employment status of their breadwinners.

12. Economically active persons and economically inactive members of their families in the private and public sectors of the economy by major industrial group.

13. Economically active persons and economically inactive members of their families by individual occupation, employment status, sex, origin, and marital status.

14. Economically active population by individual occupation, employment status, sex, and age.

15. Members of families working together by major industrial group and employment status.

16. Married women by year of marriage, number of live-born children, and age at marriage.

PAGE LOCATION MATRIX

Volume Number		Table Number															
		1	2	3	4	5	6	7	8*	9	9a	10	11	12	13	14	15	16
2.	Kanton Aargau	49	52	62	72	82	85	86	90	98	103	104	114	116	122	146	155	156
3.	Kant. Appenz. Aus.	44	46	48	50	52	53	54	58	60	65	66	76	78	82	102	105	106
4.	Kant. Appenz. Inn.	45	46	46	48	50	51	52	56	58	63	64	74	76	82	88	100	102
5.	Kant. Basel-Land.	46	48	52	56	60	64	66	70	78	83	84	94	96	102	126	131	132
6.	Kant. Basel-Stadt	44	46	46	46	48	50	52	56	60	65	66	76	78	84	108	114	116
	Gemeinde Basel	--	--	--	--	--	51	--	--	--	--	--	--	--	--	--	--	--
	Agglom. Basel	--	--	--	--	--	--	--	--	120	125	126	136	138	144	168	175	--
7.	Kanton Bern	57	62	82	102	122	128	132	144	162	167	168	178	180	186	210	222	224
	Gemeinde Bern	--	--	--	--	--	--	--	--	232	237	238	248	250	256	278	284	226
	Gemeinde Biel	--	--	--	--	--	--	--	--	286	291	292	302	304	310	332	335	228
8.	Canton Fribourg	50	52	64	76	88	90	92	96	100	105	106	116	118	124	148	153	154
9.	Canton Genève	44	46	48	50	52	54	56	64	68	73	74	84	86	92	118	126	128
	Commune Genève	--	--	--	--	--	55	--	--	134	139	140	150	152	158	184	191	--
10.	Kanton Glarus	44	46	48	50	52	53	54	58	60	65	66	76	78	84	104	107	108
11.	Kant. Graubünden	50	52	64	74	86	87	88	92	94	99	100	110	112	118	140	145	146
12.	Kanton Luzern	46	48	52	56	60	63	66	74	80	85	86	96	98	104	128	135	136
	Gemeinde Luzern	--	--	--	--	--	64	--	--	142	147	148	158	160	166	188	191	--

*Table 8 furnishes the indicated data for the canton named in the volume title, and it may also furnish data for principal communes and agglomerations within the canton.

PAGE LOCATION MATRIX (continued)

Volume Number		1	2	3	4	5	6	7	8*	9	9a	10	11	12	13	14	15	16
13.	Canton Neuchâtel	47	50	54	58	62	65	66	70	76	81	82	92	94	100	126	132	134
14.	Kanton St. Gallen	46	48	52	56	60	63	66	74	80	85	86	96	98	104	128	136	138
	Gemeinde St. G.	--	--	--	--	--	64	--	--	144	149	150	160	162	168	190	194	--
15.	Kant. Schaffhausen	44	46	48	50	52	54	56	60	64	69	70	80	82	88	110	113	114
16.	Kanton Schwyz	44	46	48	50	52	53	54	58	60	65	66	76	78	84	106	109	110
17.	Kanton Solothurn	47	50	56	62	68	72	74	78	86	91	92	102	104	110	134	141	142
18.	Kanton Thurgau	50	52	62	72	82	85	86	90	96	101	102	112	114	120	142	148	150
19.	Cantone Ticino	48	50	60	70	80	84	86	90	98	103	104	114	116	122	146	153	154
20.	Kant. Nidwalden	44	46	46	48	50	51	52	56	58	63	64	74	76	82	100	103	104
21.	Kant. Obwalden	44	46	46	48	50	51	52	56	58	63	64	74	76	82	100	103	104
22.	Kanton Uri	44	46	48	50	52	53	54	58	60	65	66	76	78	84	102	105	106
23.	Canton Valais	48	50	58	66	74	75	76	80	82	87	88	98	100	106	130	135	136
24.	Canton Vaud	52	56	72	88	104	110	112	120	132	137	138	148	150	156	182	192	194
	Commune Lausanne	--	--	--	--	--	111	--	--	200	205	206	216	218	224	248	254	196
25.	Kanton Zug	44	46	46	48	50	51	52	56	58	63	64	74	76	82	102	104	106
26.	Kanton Zürich	48	52	60	68	76	83	86	98	114	119	120	130	132	138	162	174	176
	Gemeinde Zürich	--	--	--	--	--	--	--	--	184	189	190	200	202	208	232	241	178
	Gem. Winterbur	--	--	--	--	--	--	--	--	244	249	250	260	262	268	290	293	180

SWITZERLAND

VOLUME 27. SWITZERLAND. PART I. SEX, ORIGIN, BIRTHPLACE, RELIGION, MOTHER TONGUE, MARITAL STATUS, AGE, EDUCATIONAL LEVEL.

TABLES

Number	Title	Page
	Principal Results of Population Censuses of 1850-1960	
	Occupied Houses and Households since 1860; Population since 1850	
1.	Occupied houses since 1860, cantons.	20
2.	Households since 1860, cantons.	21
3.	Resident population since 1850, cantons.	22
4.	Average population increase since 1850, cantons.	23
5.	Components of change in the Swiss population since 1870, and in the cantons since 1950.	24
	Origin	
6.	Resident Swiss population by class of origin (nationality) since 1850.	25
7.	Resident population of the cantons by class of origin (nationality) since 1860.	26
8.	Swiss citizens by canton of origin since 1850.	30
9.	Foreigners by country of origin since 1860.	31
	Birthplace[1]	
10.	Resident Swiss population by birthplace class,[2] sex, and origin since 1860.	32
11.	Resident Swiss population by canton of birth since 1888.	34
12.	Resident Swiss population by religion, sex, and origin since 1860.	36

---

[1] The birthplace is the usual place of residence of the parents or the mother when the child was born. For people born outside Switzerland, the place of birth is in the country determined by present-day boundaries.

[2] Birthplace classes of people born in Switzerland are: (1) birthplace in current commune of residence, (2) birthplace in another commune of the current canton of residence, and (3) birthplace in another canton.

13.	Resident population of cantons by religion since 1860.	38
14.	Religion of couples living together since 1880.	42

Mother Tongue

15.	Resident Swiss population by mother tongue, sex, and origin since 1880.	44
16.	Resident population of cantons by mother tongue since 1880.	46

Marital Status

17.	Resident Swiss population by marital status, sex, and origin since 1860.	48
18.	Swiss population of marriageable age by sex, marital status, and origin since 1860.	50
19.	Resident population of cantons by sex and marital status since 1860.	52
20.	Foreigners by canton of residence, marital status, and sex since 1900.	58

Age

21.	Resident Swiss population by age and sex since 1860.	64
22.	Resident Swiss population by five-year age group since 1860.	66
23.	Resident Swiss population by five-year age group and marital status since 1860.	68
24.	Proportions of each sex of the resident Swiss population by five-year age group since 1900.	70

Results of the General Population Census of December 1, 1960

25.	General table.	72

Origin

26.	Resident population of cantons by class of origin (nationality) and sex.	76
27.	Swiss citizens by canton of residence and canton of origin.	82
28.	Foreigners by country of origin and sex.	83
29.	Foreigners by country of origin, sex, and canton of residence.	86

Birthplace[1]

30.	Resident population of cantons by birthplace class,[2] origin, and sex.	90

SWITZERLAND

31.	Persons born in Switzerland by canton of residence and canton of birth.	92
32.	Persons born in a foreign country by country of birth, origin, and sex.	93

Religion

33.	Resident population of cantons by religion, origin, and sex.	98
34.	Persons of "other religion or without religion" by canton of residence, origin, religion, and sex.	102

Mother Tongue

35.	Resident population of cantons by mother tongue, origin, and sex.	104
36.	Persons of foreign mother tongue by canton of residence, origin, language, and sex.	106

Marital Status

37.	Resident population by canton, commune size-class, origin, marital status, and sex.	108

Age

38.	Resident Swiss population by year of birth, origin, marital status, and sex.	116
39.	Resident Swiss population by five-year age group, origin, marital status, and sex.	128
40.	Resident population of cantons by five-year age group and sex.	134

Couples, Married Women

41.	Married women in Switzerland by year of marriage, number of children born alive within the present marriage, and age at marriage.
	    Swiss
	    Foreigners
	    Economically active women.
		148
		150
		152
		154

42.	Married women in cantons and towns by number of children born alive within the present marriage.
	    Swiss
	    Foreigners
	    Economically active women.
		156
		157
		158
		159

43.	Religion of couples living together by canton of residence and origin.	160
44.	Mother tongue of couples living together by canton of residence and origin.	164
45.	Swiss women married to foreigners by canton of residence and age.	168

46.	Age of married persons by age of spouse among couples living together.	170
47.	Age difference of couples living together.	176

Education

48.	Persons attending school by canton of residence, origin, sex, level of education, and age group.	180
49.	Persons no longer attending school by age group, sex, and last (continuous) school.	186

VOLUME 28. SWITZERLAND. PART II. INDUSTRY AND OCCUPATIONS.

TABLES

Number	Title	Page
	Principal Results of Population Censuses of 1850-1960	
	Resident Population by Economic Dependence since 1888	
1.	Total population.	17
2.	Men.	18
3.	Women.	19
4.	Resident population by major industrial group and economic dependence since 1888.	20
5.	Economically active Swiss population by major industrial group and employment status since 1900.	
	Total	24
	Men	28
	Women	32
6.	Economically active foreign population by major industrial group and employment status since 1900.	
	Total	36
	Men	40
	Women	44
7.	Economically active non-independent persons in private and public sectors of the economy by major industrial group since 1930.	48
8.	Economically active persons by certain occupations and employment status since 1941.	50
9.	Agricultural population since 1888.	54
	Results of the General Population Census of December 1, 1960	
10.	General table.	56
11.	General table by canton.	58

SWITZERLAND

12.	Economically active persons and economically inactive members of their family by industrial group.	64
12a.	Staff, economically inactive members of their family, and inmates of institutions by kind of institution.	69
12b.	Economically inactive independent persons and economically inactive members of their family.	70
12c.	Economically inactive persons living in a family other than their own.	71
13.	Economically active population by industrial group, marital status, origin, and employment status.	72
14.	Economically inactive members of the family of economically active persons by major industrial group, sex, marital status, origin, and employment status of their supporters.	120
15.	Independent economically active persons by major industrial group and type of their employees and workers.	134
16.	Cooperating family members by major industrial group and employment status.	135
17.	Economically active persons and economically inactive members of their family in private and public sectors of the economy by major industrial group.	136
18.	Economically active population by industrial group, employment status, and age.	144
19.	Economically active women and economically inactive independent women by economic dependence, industrial division, marital status, age group, and number of family members dependent on her and living in the same household.	160
19a.	Economically active married women and economically inactive independent married women by economic dependence, industrial division, marital status, age group, and number of family members dependent on her and living in the same household.	162
20.	Economically active population by individual occupation, marital status, employment status, origin, and sex.	164
21.	Economically inactive members of families of economically active persons by individual occupation, employment status, origin, and sex of their supporters.	222
22.	Economically active population by individual occupation, employment status, marital status, origin, sex, and age.	243

23.	Economically active population by major industrial group and individual occupation.	283
24.	Persons having learned an occupation by economic dependence and sex.	302
25.	Economically active persons by occupation learned, occupation practiced, employment status in occupation practiced, and sex.	313
26.	Persons having a secondary occupation[1] by industrial division and sex.	342
27.	Persons having a secondary occupation[1] by their principal and secondary occupation.	350
28.	Economically inactive population by origin, age, and sex.	373
29.	Housewives by secondary occupation,[1] marital status, and age.	374
30.	Resident population by socio-economic group, sex, and age.	376
31.	Economically active persons and economically inactive independent persons, together with the economically inactive members of their families, by socio-economic group, sex, and origin.	377

VOLUME 30. SWITZERLAND. PART IV. RESIDENCE AND WORKPLACE OF ECONOMICALLY ACTIVE POPULATION (COMMUTERS).[2]

TABLES

Number	Title	Page
1.	Daily commuters into and out of cantons and towns in 1960.	26

---

[1] A secondary occupation is an occupation practiced for some time, and which yields more than a temporary or modest increase to principal income. Employment status is the same as for main occupation. Classification of secondary occupation is by industry, except where an activity cannot be attached to a given industry.

[2] This volume covers only those persons who work outside their commune of residence and return home every day. It does not cover people moving within the commune, persons who reside in a foreign country and work in Switzerland, or students. The working population comprises the economically active population of the commune minus the number of outgoing migrants plus the number of incoming migrants.

SWITZERLAND

2.	Resident population and working population by canton and town in 1960.	27
3.	Inter-cantonal commuting in 1960.	28
4.	Commuting economically active persons by commune in 1960.	31
5.	Incoming economically active commuters by certain communes of residence in 1960.	61
6.	Outgoing economically active commuters by certain communes of work in 1960.	83
7.	Economically active commuters by demographic and occupational characteristics in 1960: Incoming commuters Outgoing commuters	94 96

VOLUME 31. SWITZERLAND. PART V. COMMUNES OF ORIGIN OF SWISS CITIZENS.[1]

TABLES

Number	Title	Page
1.	Swiss citizens by canton of origin and place of residence since 1860, first communal citizenship. Absolute numbers.	8
1a.	Swiss citizens by canton of origin and place of residence since 1850, first communal citizenship. Proportional numbers.	10
2.	Swiss citizens residing in Switzerland by canton of origin and residence in 1960, first communal citizenship.	12
3.	Swiss citizens residing in Switzerland by commune of origin and canton of residence in 1960, first communal citizenship.	21
4.	Swiss citizens residing in Switzerland by commune of origin and canton of residence in 1960, all communal citizenships.	21

---

[1] Citizenship in the Republic of Switzerland derives from prior citizenship in a commune. A first communal citizenship (Bürgerrecht, droit de cité) is inherited from one's father and is retained throughout life. Additional communal citizenships can be obtained by moving to and residing in a new commune for a period of time determined by the commune. Therefore, a person's communal citizenship and commune of residence may not refer to the same commune.

5.	Swiss citizens residing in Switzerland by canton of origin and place of residence in 1960, all communal citizenships.	22
6.	Communal citizenships of Swiss citizens living in Switzerland by canton of origin and residence in 1960, all communal citizenships.	80
7.	Swiss citizens living in Switzerland by canton of residence and number of their communal citizenships in 1960.	82
8.	Citizens originating from a single canton and residing in Switzerland by canton of origin and the number of their communal citizenships in 1960.	83
9.	Citizens originating from two cantons and living in Switzerland by canton of origin in 1960.	84
10.	Citizens originating from more than two cantons living in Switzerland in 1960.	86

VOLUME 32. SWITZERLAND. PART VI. HOUSEHOLDS.

TABLES

Number	Title	Page
1.	Distribution of households (including institutional households) by type, cantons and towns.	19
2.	Private households by type and size.	20
3.	Private households by type and size by canton.	24
3a.	Private households by type and size, towns and population aggregates.	32
4.	Private households by age and sex of the head of the household and by type of household.	37
5.	Persons living in private households by their status in the household, marital status, sex, and origin.	41
6.	Persons living in family households by their status in the household, cantons and towns.	42
7.	Persons living in non-family households by their status in the household, cantons and towns.	45
8.	Family households by marital status and sex of the head of the household, cantons and towns.	46
9.	Non-family households by marital status and sex of the head of the household, cantons and towns.	48
10.	Family households including children of the head of the household by type of household, number of children, and sex and age of the head of the household.	50

SWITZERLAND

11.	Family households by number of children of the head of the household.	52
12.	Spouses living together of different religious denominations by religion adopted for their children.	53
13.	Family households including children of the head of the household who are less than 16 years of age by mother tongue of parents and children.	54
14.	Spouses living together and their children by language and religion of the couple.	56
15.	Non-family households of one person by religion, language, origin, marital status, and sex.	59
16.	Family households of spouses living together by nationality of the couple and by unmarried children of the head of the household.	60
17.	Family households by age and educational level of the head of the household, size of household, and number of persons less than 16 years of age.	62
18.	Family households with children and students by number and age of both and by educational level they have obtained.	64
19.	Heads of households attending school by economic dependence, marital status, sex, age, and number of children.	67
20.	Private households by type of household and number of employees belonging to the household.	70
21.	Private households by type of household and number of lodgers, cantons and towns.	71
22.	Institutional households consisting of persons 65 years of age and older.	72
23.	Private households by size, socio-economic group, and sex of the head of the household.	75
24.	Private households by type, socio-economic group, and sex of the head of the household.	76
25.	One-couple households by socio-economic group, age of the husband, and occupation of the wife.	78
26.	Family households of couples not having lodgers by socio-economic group of the husband and economic activity of the wife.	80
27.	Non-family households comprising only "other related persons" by socio-economic group and by sex of the head of the household.	81
28.	Private households by type, size, and number of persons who are economically active.	83

29.	Private households in which the head of the household is economically active by household size, industrial division, and sex of the head of the household.	84
30.	Private households by size and individual occupation and sex of the head of the household.	86
31.	Family households by economic activity, industrial division, and employment status of the head of the household, and by economic activity of persons less than 16 years of age.	90
32.	Family households by economic activity and employment status of the head of the household, economic activity of children over 16 years of age, and number of children less than 16 years of age.	92
33.	Family households comprised of economically active persons by their economic activity in agriculture and by size of the household.	96
34.	Family households managed by a couple by age and economic activity of the husband and number of children less than 7 years of age supported by the couple, for cantons, towns, and population aggregates.	98
35.	Private households by number and major industrial group of the craftsman's helpers.	106

Households and Dwellings

36.	Family and non-family households by number of persons, size, and conveniences of the dwellings.	110
37.	Households by composition and by size and conveniences of the dwellings.	112
38.	Family households of couples by sex and number of children and by size and conveniences of the dwellings.	114
39.	Family households of couples having children by age, sex, and number of children, and by size and conveniences of the dwellings.	115
40.	One-person family households having children by age, sex, and number of children, and by size and conveniences of the dwellings.	122

APPENDIX A

ECONOMIC ACTIVITY

The following is an abbreviated and edited quotation from the recommendations of the European Conference of Statisticians, found in European Population Censuses: The 1960 Series, pp. 7-12:

The classification of the population by economic activity designates all persons as either economically active or inactive.

The distinction between the economically active and inactive population should be based on the length of time worked. The criterion of (actually or presumably) receiving an income should be used only to distinguish the independents from the dependents within the economically inactive population.[1]

Economically active population: This group comprises all persons of either sex who furnish the supply of labor available for the production of economic goods and services. It includes both employed and unemployed persons during the time reference period adopted for the census. Members of the armed forces should be included in the economically active population. They should be shown as a separate subgroup, or their position should be clearly described in the census report. Optionally the employed and the unemployed may be shown separately in the census.

The employed comprise all persons, including family workers, who are at work or who have jobs during the reference period, whether they are full-time or part-time workers, provided that the latter work at least a minimum period to be set by each country, sufficiently long to exclude those whose contributions are negligible.

---

[1] In actual practice, all countries in this volume (with the exception of Portugal) subdivided the economically inactive population into independents and dependents. However, countries differed with respect to the coverage and definition of those included in these categories. The reader interested in such detailed information may refer to Table 10, "Data on Type of Activity," pp. 119-122, of European Population Censuses: The 1960 Series (European Conference of Statisticians, Statistical Standards and Studies, No. 3).

APPENDIX A

The <u>unemployed</u> consist of all persons above a specific age who during the reference period are not working but who are seeking work for pay or profit. Persons seeking their first job should be included. Their position should be made clear either by showing them as a separate subgroup or by textual explanations. Persons who are not seeking work in the reference period because of minor illness, because they have made arrangements for a new job, or because they are on temporary or indefinite layoff without pay, are also included.

<u>Economically inactive population</u>: This group comprises all persons of either sex who are not economically active. It may be divided into 2 subgroups: the independent and the dependent.

<u>Independent persons</u> are those whose principal source of livelihood is, or is presumed to be, a personal income.

   Optional subgroups:
     a. Former members of the economically active population
     b. Students having a personal income
     c. Other independent persons (e.g., rentiers)

<u>Dependent persons</u> are all others.

   Optional subgroups:
     a. Children below the minimum school-leaving age
         1. Children below school-commencing age
         2. Children of school age
     b. Other students (i.e., those without personal income)
     c. Homemakers (e.g., housewives and other relatives)
     d. Persons in institutions
     e. Others

APPENDIX B

EMPLOYMENT STATUS

The following is a paraphrase of the definition appearing in Population Census Methods (New York: United Nations Population Studies No. 4, 1949), p. 116:

"Employment status" (sometimes called "occupational status," "class of worker," "social status," or "position in industry") refers to the classification distinguishing employees, employers, workers on own account, family workers, and similar categories.

The following classification by employment status was recommended by the European Conference of Statisticians in European Population Censuses: The 1960 Series, p. 20:

I. Employers
II. Own-account (independent) workers
III. Employees

   Optional subgroups

   (i) Apprentices
   (ii) Other possible groups

IV. Family workers

   Optional subgroups

   (i) Paid family workers
   (ii) Unpaid family workers

V. Members of producers' cooperatives
VI. Economically active persons not classifiable by status
VII. Independent inactive persons

In this classification unemployed persons should be classified according to their former employment status.

APPENDIX B

Similarly, groups may be further subdivided into subgroups, and the ISIC provides examples of possible subgroups 611 and 612 of the industrial classification.

Some examples of the international industrial classification are given below:

Division	Major Group	Group	Subgroup	
6				Commerce
	61			Wholesale and retail trade
		611		Wholesale trade
			6111	Agricultural raw materials
			6112	Minerals, metals, etc.
			6113	Lumber and construction materials
		612		Retail trade
			6121	Grocery, food, liquor stores
			6122	Pharmacies, drug stores
			6123	Dry goods, apparel, footwear stores
7				Transport, Storage, and Communication
	71			Transport
		711		Railway transport
		712		Tramway and omnibus operators
		etc.		
	72	720		Storage and warehousing
	73	730		Communication

APPENDIX C

INDUSTRIAL CLASSIFICATION

The following is an abbreviated and edited quotation from the <u>International Standard Industrial Classification of All Economic Activities</u> (United Nations Statistical Papers, Series M, No. 4, Rev. 1), pp. 2-5, 26, 27. This source is referred to throughout the glossaries as "ISIC 1958":

The ISIC 1958 is a classification by kind of industry and not by kind of occupation or commodity. The classification does not draw distinctions according to kind of ownership, type of economic organization, or mode of operation. Thus establishments engaged in the same kind of economic activity are classified in the same group of the ISIC, irrespective of whether they are owned by incorporated enterprises, individual proprietors, or governments, or whether or not the parent enterprise owns other establishments. Similarly, manufacturing establishments are classified according to the kind of economic activity in which they engage, whether the work is performed by power-driven machinery or by hand, or whether it is done in a factory or a household.

The complete industrial classification system comprises nine divisions. Each division is assigned its permanent one-digit number, except manufacturing which, because of the number of major groups separately recognized, receives two one-digit numbers (2 and 3). Each division has 10 subdivisions called major groups, and these are identified by two-digit numbers. The first digit indicates the division, and the first and second digits taken together identify the major groups comprising the division. Each major group, in turn, may be divided into groups. The industrial classification is thus read as follows: The first and second digits taken together indicate the major group; and the first, second, and third digits taken together identify the group.

In cases where a major group is not subdivided into groups, the title of the major group is also used as the title of the three-digit group with "0" added to the identification number of the major group. For example, no further subdivisions are shown for major group 73, Communication, and the three-digit number is therefore 730.

APPENDIX C

    9   Service, sport, and recreation workers
    X   Workers not classifiable by occupation
    --   Members of the armed forces

These major groups are divided into 73 minor groups, which in turn are made up of 201 unit groups. The major, minor, and unit groups constitute the international classification endorsed by the Ninth International Conference of Labor Statisticians. Finally, unit groups are subdivided into occupations, of which there are a total of 1,345.

The decimal coding system employed facilitates understanding of the general structure of the classification and reflects the relationships between occupations and groups of occupations. The assignment of code numbers is as follows:

  Major groups: Major groups are identified by the initial digit in the code number; major group 7/8 utilizes two initial digits.

  Minor groups: Minor groups are identified by code numbers of two digits separated by a hyphen (-). The first digit indicates the major group in which the minor group falls. Occasionally, the eleventh and twelfth minor groups terminating major group 0 (minor groups 0-X and 0-Y) are identified by letters.

  Unit groups: Unit groups are identified by code numbers of three digits, the first two representing the minor group to which the unit group belongs.

  Occupations: Occupations are identified by code numbers of five digits, the first three representing the unit group in which the occupation is included. A point (.) separates the final two digits from the unit group code number.

Examples: Five-digit (individual occupation) designations

8-12 POTTERS AND RELATED CLAY AND ABRASIVE FORMERS

   8-12.10 Potter
   8-12.15 Modeller, Pottery and Porcelain
   8-12.20 Mould Maker, Pottery and Porcelain
   8-12.25 Thrower, Pottery and Porcelain

APPENDIX D

OCCUPATIONAL CLASSIFICATION

The following is an abbreviated and edited quotation from the International Standard Classification of Occupations (Geneva: International Labour Office, 1958), p. 4. This source is referred to throughout the glossaries as "ISCO 1958":

The ISCO 1958 classification by occupation enumerates persons by kinds of work performed (or previously performed in the case of the unemployed), irrespective of the industrial classification or employment status of those persons.

Basic Components of the System

The ISCO 1958 consists of three basic components--namely, code numbers, titles, and definitions. Since the structure of the classification is based primarily on the nature of the work done, the code number and title of an occupation or group must obviously not be considered separately from the corresponding definition, which defines the scope and content of the occupation or group concerned.

In the classification structure all civilian occupations are divided into ten major groups, including the supplementary major group X (Workers Not Classifiable by Occupation). Members of the armed forces are included in a separate group. The complete list of major groups is as follows:

Code No.	Title
0	Professional, technical, and related workers
1	Administrative, executive, and managerial workers
2	Clerical workers
3	Sales workers
4	Farmers, fishermen, hunters, loggers, and related workers
5	Miners, quarrymen, and related workers
6	Workers in transport and communications occupations
7/8	Craftsmen, production process workers, and laborers not elsewhere classifiable

APPENDIX E

SOCIO-ECONOMIC CATEGORIES

The following is an abbreviated and edited quotation from the recommendations of the European Conference of Statisticians, found in <u>European Population Censuses: The 1960 Series</u>, pp. 22-24:

The classification by socio-economic category identifies different population groups which are, on the one hand, reasonably homogeneous, and on the other, fairly clearly distinguishable from other groups in behavior. It can therefore be used to establish the relationship between the socio-economic position of individuals (and households) and many demographic, social, economic, and cultural phenomena.

The socio-economic classification is basically a derived classification, not requiring additional questions on the census questionnaire but deriving from other classifications—namely, type of economic activity, occupation, industry, and employment status (as employer, employee, etc.).

Some examples of the 26 recommended categories are:

<u>Economically active population</u>

(1) Farmer-employers

(2) Farmers on own account without employees

(3) Members of agricultural producers' cooperatives

(4) Agricultural workers

(5) Employers in industry and commerce: large enterprises

(6) Employers in industry and commerce: small enterprises

<u>Economically inactive population</u>

(18) Former farmer-employers

(19) Former non-agricultural employers

(20) Former employees

APPENDIX F

HOUSEHOLDS AND FAMILY NUCLEI

The following is an abbreviated and edited quotation from the recommendations of the European Conference of Statisticians found in European Population Censuses: The 1960 Series, pp. 7-12:

A clear distinction should be made between the concepts of "household" and "family." The household is a basic population census concept which serves first as an important unit of enumeration and, subsequently, as a general framework within which census statistics on households and families can be obtained. While the household is identified by the census enumerator, the family is determined at the processing stage by combining the information for some or all of the individual members of the family. A household can consist of more than one family, but a family in the sense of a family nucleus, as defined below, cannot be composed of two or more households. A family always constitutes a private household or part of a larger private (or institutional) household. In practice, the private household and the family frequently coincide.

Households

The household definition should distinguish two broad classes: private and institutional.

A private household should preferably be defined as:

a. A one-person household where one person lives alone in a separate housing unit or occupies, as a lodger, part of a separate room or rooms in part of a housing unit, but who does not join with any of the other occupants of the housing to form part of a multi-person household as defined below;

or

b. A multi-person household where a group of two or more persons together occupy the whole or part of a housing unit and provide themselves with food or other essentials for living. The group may be composed of related persons only, of unrelated persons, or of a combination of both, including boarders but excluding lodgers.

APPENDIX F

This concept of household might for convenience be referred to as the "housekeeping unit" concept. However, in some countries it is the practice to use a different concept which equates the household with the housing unit and defines the household as the entire group of persons jointly occupying a housing unit. This concept of household (which might be referred to as the "household-dwelling" concept) does not provide direct information on the number of housekeeping units sharing housing units and should be avoided unless the number of housing units actually inhabited by two or more housekeeping units is very small, i.e., where in the particular circumstances of the country, the household (housekeeping unit) is generally co-terminus with the housing unit. In this connection, it is important to bear in mind that housing units and households, while they are interdependent in the sense that one should not be considered without reference to the other, are clearly distinguishable concepts.

<u>Institutional households</u> comprise groups of people living in hotels, boarding houses, colleges, schools, hospitals, etc., who are subject to common authority and are bound by a common public or private objective and/or personal interests or characteristics. In addition to hotels and boarding houses so described, households in which the number of boarders exceeds five should be enumerated as institutional households. The households of institutional directors and administrative personnel with separate living quarters should be considered as private households.

Classification of institutional households should be based on the main purpose which they are designed to serve, as follows:

a. Residential--hotels, boarding houses, and lodging houses
b. Educational--colleges and schools
c. Medical--general hospitals, mental hospitals, sanatoria, and convalescent homes
d. Welfare--homes for the aged, homes for unmarried mothers, and orphanages
e. Religious--those monasteries and convents which are not classified as educational, medical, or welfare institutions
f. Military--barracks, forts, and naval vessels
g. Penal--prisons and reformatories
h. Others

Families

The <u>family</u> is defined as those persons within a household who are related by blood, marriage, or adoption. This definition includes stable <u>de facto</u> unions as well as marriage.

APPENDIX F

The analysis of families can only be precise and meaningful if it is undertaken in terms of <u>family nuclei</u>. The three types of family nuclei are as follows:

a. Married couple with one or more unmarried children
b. Married couple without children
c. One parent with one or more unmarried children

In this definition "unmarried" means "never married." A never-married woman with one or more children is treated as a separate family nucleus even if they are living in the same household as the mother's parents. "Children" include foster as well as adopted children.

In contrast to the "family nucleus" concept, the family may be viewed as all the related members of a household.

Family Classification of Private Households

<u>Private households</u> should be classified by family composition as follows:

A. Non-family households
   1. One-person
   2. Multi-person
      a. Related persons
      b. Unrelated persons
      c. Related and unrelated persons
B. One-family households
   1. Married couple with one or more unmarried children
   2. Married couple without children
   3. One parent with one or more married children
C. Multi-family households
   Multi-family households should be classified by whether or not any of the family nuclei are related and whether the relationship is by direct descent. The family nuclei should also be classified as outlined above.

Head of Household/Household Supporter

The <u>head of the household</u> is usually considered to be that person who is acknowledged as such by the other household members. It is more important, however, for the purposes of dependency statistics, to identify the person on whom falls the chief responsibility for the economic maintenance of the household, i.e., the main breadwinner or the principal contributor to the household budget, who may be called the "household supporter."

INDEX: SELECTED TOPICS TABULATED BY NATIONAL CENSUSES, 1960[a]

NATION	CULTURAL, EDUCATIONAL, AND PERSONAL CHARACTERISTICS							ECONOMIC			
	Birthplace	Citizenship	Education	Internal migration	Language	Nationality	Religion	Commuting	Economically active	Economically inactive	Employment status
Austria	O	O	X	O	X	X	X	X	X	X	X
Belgium	X	O	X	O	O	X	O	X	X	X	X
Denmark	O	O	O	O	O	O	O	O	X	O	X
Finland	X	X	X	O	X	O	X	X	X	O	X
France	O	X	X	X	O	X	O	O	X	X	X
Germany	O	X	X	O	O	X	X	X	X	X	X
Gibraltar	X	X	O	O	O	X	X	O	O	O	O
Great Britain	X	X	X	X	X	X	O	X	X	X	X
Greece	O	O	X	X	O	X	O	O	X	X	X
Ireland (Eire)	X	O	O	O	X	O	X	O	X	X	X
Italy	X	X	X	O	X	O	O	O	X	X	X
Liechtenstein	X	X	O	O	X	X	X	O	X	X	X
Luxemburg	X	X	O	O	O	X	X	X	X	X	X
Malta	O	X	X	O	O	X	O	O	X	X	X
The Netherlands	X	O	X	X	O	X	X	X	X	X	X
Northern Ireland	X	O	X	O	O	X	X	O	X	X	X
Norway	X	X	X	X	O	X	X	X	X	X	X
Portugal	O	X	X	X	O	X	X	O	X	X	X
Scotland	X	X	X	X	X	X	O	X	X	X	X
Spain	X	O	X	O	O	X	O	O	X	X	X
Sweden	X	X	X	X	O	X	O	X	X	X	X
Switzerland	X	X	X	X	X	X	X	X	X	X	X

[a]This index is based on the occurrence of technical terms in the volume title or table titles of at least one volume per complete census text. Concepts which could be assumed to be universally tabulated (e.g., sex, age) were excluded from this listing, as were concepts whose incidence (on a comparative basis among the nations) was relatively infrequent.

INDEX (continued)

CHARACTERISTICS			HOUSEHOLD AND FAMILY CHARACTERISTICS						POPULATION DEFINITIONS		POPULATION DENSITY CONCEPTS
Industrial classification	Occupational classification	Socio-economic category	Family	Fertility	Households, Heads of	Households, Institutional	Households, Private	Marital status	De facto population	De jure population	Population cluster ("locality") concepts
X	X	O	O	O	X	X	X	X	O	X	O
X	X	X	X	O	X	X	X	X	O	X	O
X	X	X	X	O	X	X	X	X	O	X	X
X	X	X	X	O	X	O	X	X	O	X	X
X	X	X	X	O	X	X	X	X	O	X	X
X	X	X	X	X	X	X	X	X	O	X	O
X	X	O	O	O	O	O	O	X	X	X	O
X	X	X	X	X	X	X	X	X	X	O	O
X	X	O	O	O	X	X	X	X	X	X	X
X	X	X	X	X	O	O	O	X	X	O	O
X	X	O	O	O	X	X	X	X	X	X	O
X	X	O	O	X	O	X	X	X	X	X	O
X	X	X	X	O	X	X	X	X	X	X	O
X	X	X	O	O	O	O	X	X	O	X	O
X	X	X	X	X	X	X	X	X	O	X	X
X	X	X	X	X	X	X	X	X	X	O	O
X	X	X	X	X	X	X	X	X	O	X	X
X	X	X	X	X	X	X	X	X	X	X	O
X	X	X	X	X	O	X	X	X	X	O	O
X	X	O	O	X	O	O	O	X	X	X	O
X	X	X	X	X	X	X	X	X	O	X	X
X	X	X	X	X	X	X	X	X	O	X	X

WITHDRAWN
UST
Libraries

O'Shaughnessy - Frey Library
University of St. Thomas
St. Paul, MN 55105

DATE DUE

HA 37 .E93 B5 1971

Blake, Judith.

Western European censuses, 1960

DEMCO